# SEM:

*A User's Manual for*
*Materials Science*

# SEM:

## A User's Manual for Materials Science

Barbra L. Gabriel

*Staff Consultant in Microscopy*
*Packer Engineering Associates, Inc.*

AMERICAN SOCIETY FOR METALS
Metals Park, Ohio 44073

Editorial and production coordination by
Carnes Publication Services, Inc.
Project manager: Nancy I. Tenney

Library of Congress Catalog Card Number: 85-71081
ISBN: 0-87170-202-9
SAN: 204-7586

PRINTED IN THE UNITED STATES OF AMERICA

*In memoriam*

ANTHONY R. GABRIEL

## ACKNOWLEDGMENTS

Sincere gratitude to my associates at Packer Engineering for their cooperation and assistance. I am particularly indebted to Mr. Charles Morin for sharing his knowledge and experience. Special thanks to Ms. Linda Maurer for her superb secretarial skills and patience.

# Contents

# PART 1:
## Instrumentation

# =1=

# The Scanning Electron Microscope

The scanning electron microscope (SEM) has unique capabilities for analyzing surfaces. It is analogous to the reflected light microscope, although different radiation sources serve to produce the required illumination. Whereas the reflected light microscope forms an image from light reflected from a sample surface, the SEM uses electrons for image formation. The different wavelengths of these radiation sources result in different resolution levels: electrons have a much shorter wavelength than light photons, and shorter wavelengths are capable of generating higher-resolution information. Enhanced resolution in turn permits higher magnification without loss of detail. The maximum magnification of the light microscope is about 2000×; beyond this level is *"empty magnification,"* or the point where increased magnification does not provide additional information. This upper magnification limit is a function of the wavelength of visible light, 2000 Å, which equals the theoretical maximum resolution of conventional light microscopes. In comparison, the wavelength of electrons is less than 0.5 Å, and theoretically the maximum magnification of electron beam instruments is beyond 800,000×. Because of instrumental parameters, practical magnification and resolution limits are ~75,000× and 40 Å in a conventional SEM.

Another difference between light and scanning electron imaging concerns the *depth of field,* defined as the ability to maintain focus across a field of view regardless of surface roughness. Human binocular vision permits observation and interpretation of depth of field in three-dimensional objects. Conventional photographs and photomicrographs are two-dimensional representations; the dimension of depth is suppressed when recording an image with a diffuse light source. In contrast, SEM micrographs maintain the three-dimensional appearance of textured surfaces, a phenomenon due to the high depth of field of scanning instruments. Depth of field is further suppressed in both macro-photography and photomicrography as magnification is increased. At $10\times$, the relative depth of field of a light microscope is about 250 $\mu$m, while that of the SEM is about 1000 $\mu$m; at $1200\times$ the depth of field of a light microscope is $\sim 0.08$ $\mu$m; at $10,000\times$, the depth of field of the SEM is 10 $\mu$m. Thus, photographers are often challenged to record rough surfaces while maintaining depth of field (through determination of the "optimal aperture"); scanning electron microscopists readily record smooth or rough surfaces.

The combination of high resolution, an extensive magnification range, and high depth of field makes the SEM uniquely suited for the study of surfaces. As such, it is an indispensable tool in materials science research and development. Microscope instrumentation and operation are described in this chapter, and subsequent chapters cover energy-dispersive spectroscopy, sample preparation, and applications of SEM in materials science.

## SEM INSTRUMENTATION

The SEM (Fig. 1-1 and 1-2) consists basically of four systems:

1. The *illuminating/imaging system* produces the electron beam and directs it onto the sample.
2. The *information system* includes the data released by the sample during electron bombardment and detectors which discriminate among and analyze these information signals.
3. The *display system* consists of one or two cathode-ray tubes for observing and photographing the surface of interest.
4. The *vacuum system* removes gases from the microscope column which would otherwise interfere with high-resolution imaging.

Each of these systems and their relationships are discussed below. In addition, both the theoretical and practical aspects of microscope operation are analyzed.

## ILLUMINATING/IMAGING SYSTEM

The illuminating/imaging system comprises an electron gun and several magnetic lenses that serve to produce a collimated, coherent beam of electrons which can be focused onto the specimen (Fig. 1-3).

### Electron Gun

The electron gun can be subdivided into (1) a *filament* (cathode) or electron source, which generates electrons and is thus held at a negative potential with respect to ground; (2) an apertured *shield* at slightly

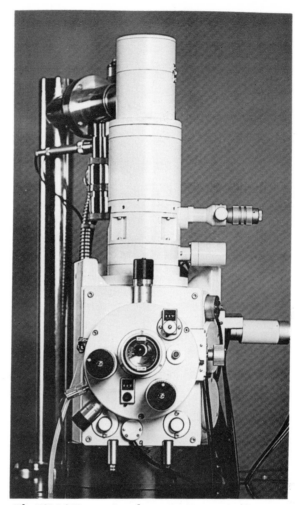

Fig. 1-1.   The JSM-840 scanning electron microscope. (Courtesy of JEOL)

positive potential relative to the filament; and (3) an *anode* held at very high positive potential with respect to the filament. Together these components function as an electrostatic lens: electrons are produced by passing a current through the filament and heating it to a point where the voltage gradient between the filament and anode produces electrons, which are then accelerated by the potential difference between the anode and filament.

The *tungsten hairpin filament* is the most common type of electron source in use today. Although analogous to incandescent light bulb filaments, SEM filaments are designed to carry a much higher voltage (10,000-20,000 volts) than light bulbs. A 0.125-mm tungsten wire is shaped into a hairpin and soldered to two electrodes which are connected to the high-voltage system (Fig. 1-3). A ceramic insulator prevents arcing between the filament and high-voltage source. Tungsten

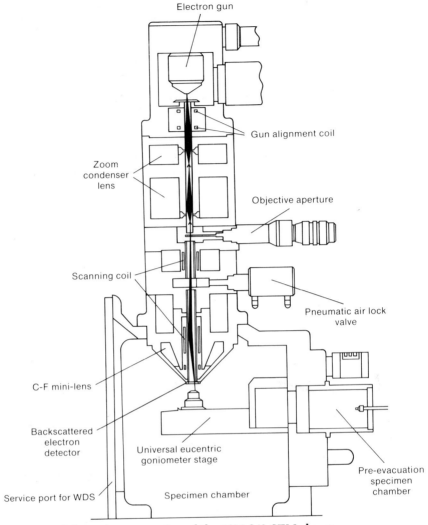

**Fig. 1-2.** Schematic cross section of the JSM-840 SEM shown in Fig. 1-1. (Courtesy of JEOL)

filaments are extremely popular because they operate under high (as opposed to ultrahigh) vacuum ($\geq 10^{-4}$ torr), they are inexpensive, and they are easy to exchange. In a well-operated microscope, tungsten filaments will function for ~40 hr before failing.

Nonconventional electron sources include *lanthanum hexaboride* (LaB$_6$) and *field emission* (FE) guns, both of which are much brighter than tungsten cathodes. The typical operating brightness for each source is as follows: tungsten hairpin, $5 \times 10^4$ to $10^5$ A/cm$^2$ steradian; LaB$_6$, $6 \times 10^6$ A/cm$^2$ steradian; FE, $10^7$ to $2 \times 10^8$ A/cm$^2$ (Table 1-1). Both types of sources, field emission in particular, offer advantages in high-resolution SEM, but they have not replaced tungsten filaments because modification of the SEM vacuum system is required for ultrahigh vacuum, special care must be exercised in operation and maintenance of

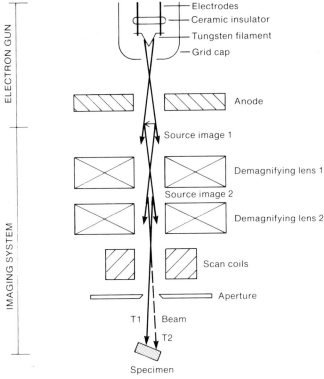

ELECTRON GUN

IMAGING SYSTEM

Electrodes
Ceramic insulator
Tungsten filament
Grid cap

Anode

Source image 1

Demagnifying lens 1

Source image 2

Demagnifying lens 2

Scan coils

Aperture

T1   Beam

T2

Specimen

**Fig. 1-3.  Cross section of the illuminating/imaging system.**

these guns, and both are expensive options. Table 1-1 summarizes the characteristics of these cathodes; further information on FE and $LaB_6$ guns may be found in Broers (1975).

Returning to our discussion of the conventional electron gun, surrounding the tungsten filament is the *shield* (synonyms: grid cap, Wehnelt cylinder). The shield is a slightly biased cylindrical cup which serves to collimate the electrons from the filament and direct them toward the anode. The round opening of the shield is centered over the filament tip. The distance separating the tip and shield (1-2 mm) is critical in that it controls beam current. A *bias* voltage is applied to the shield which allows only those electrons emitted from the tip of the filament to form the beam. If the gap between the filament tip and the grid cap is too small, the filament will quickly burn out; if the two

**Table 1-1.  Characteristics of Electron Guns (Adapted from Broers, 1975)**

| Cathode | Vacuum requirement, torr | Brightness, A/cm$^2$ steradian | Minimum beam diameter, Å | Lifetime, hr |
|---|---|---|---|---|
| Tungsten hairpin . . . . . . . . . | $>10^{-4}$ | $5 \times 10^4$ to $10^5$ | $\cong 50$ | 40-60 |
| Lanthanum hexaboride. . . . . | $>10^{-5}$ | $6 \times 10^6$ | $\cong 25$ | $\cong 3000$ |
| Field emission . . . . . . . . . . . | $>10^{-9}$ | $10^7$ to $2 \times 10^8$ | $\leq 10$ | Indefinite |

are widely spaced, the beam current is reduced and poor imaging results. Therefore, manufacturer specifications for the size of this gap must be followed for optimal operation.

Electrons passing through the aperture of the shield are attracted toward the *anode,* the third component of the electron gun. The anode is at a large potential difference relative to the filament, causing acceleration of the electrons. The difference in potential between the filament and the anode is the *accelerating voltage*. A range of voltages between 1 and 30 keV is available on most SEMs. As will be discussed later, the choice of accelerating voltage depends upon specimen type (conductive vs nonconductive samples) and the type of information desired in an analysis.

As mentioned above, the electron gun acts like an electrostatic lens to produce the imaging beam. The effective operating conditions for the electron gun are based upon incandescent heating of the tungsten filament. In this context, filament temperature and current are essentially identical, in that increasing temperature increases current. At very low filament currents, selective crystalline areas of the filament emit electrons, thereby forming a multiple electron source; this condition is referred to as *undersaturation*. As higher currents pass through the filament, the emitting areas correspondingly increase until the tip of the filament symmetrically generates electrons; this condition, which is referred to as *saturation*, is the effective electron source for imaging. Additional current input has no effect on beam intensity but drastically reduces filament lifetime (Fig. 1-4). Proper use of the SEM entails operating at saturation. Attempting operation at undersaturation reduces image clarity because electron-optical alignment cannot be achieved (having too many electron sources results in weak, multiple beam formation), whereas oversaturation drastically reduces filament lifetime. Under acceptable operating conditions, a tungsten filament will emit for ~40 hr; an abused filament will fail after only a few hours of use.

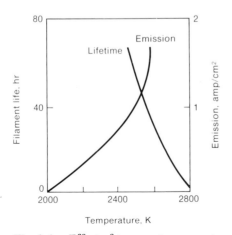

**Fig. 1-4.    Effect of temperature on a tungsten filament.**

Oxidation of tungsten, a second cause of premature filament failure, results from operating under inadequate vacuum and/or admitting air to the microscope column before the filament has cooled. Tungsten will react with any residual gas, and the reaction is accelerated if the filament is warm. The reaction site is a weak zone that cannot tolerate the stress associated with high operating currents, and consequently the filament will break at these zones, leading to zero emission. The minimum vacuum required for SEM operation is $10^{-4}$ torr: therefore, the operator must wait until high vacuum has been established before energizing the filament. A similar waiting period is necessary before admitting air to the column after the filament has been turned off (e.g., to change a specimen). The effective operating temperature for electron emission is ~2300 K (Fig. 1-4); the filament requires 3-5 min of cooling while under vacuum to minimize oxidation. Some instruments are designed with a vacuum lock at the specimen chamber; when samples are changed, only the specimen chamber is brought to atmospheric pressure. On these SEMs, it is not necessary to wait very long to exchange specimens, but after the exchange one must wait for high vacuum to be re-established.

To summarize, observing a few simple rules will take full advantage of filament lifetime:

1. Follow manufacturer specifications for the spacing between the filament tip and shield.
2. Operate at saturation, not above or below.
3. Do not turn on the filament before reaching high vacuum.
4. Cool the filament for 3-5 min before bringing the column to atmospheric pressure.

Despite these precautions, filaments have a finite lifespan (40-60 hr) and must be replaced regularly. The filament has failed when emission meters do not rise as current is increased. (Note that if current is registered but there is no image, the column is misaligned.) The general procedure for changing a filament is as follows:

1. Wearing lint-free cotton or nylon gloves, remove the gun cartridge (filament plus shield). This unit will be hot if the SEM was in the operation mode just prior to failure.
2. Disassemble the gun cartridge. The shield usually has a blue or black discoloration resulting from hydrocarbon contamination and tungsten evaporation. To minimize electron optical aberrations, the shield must be cleaned as follows:
   a. Using a metal polish (e.g., Metalputz), buff the shield until all of the discoloration is removed. Pay particular attention to the shield aperture.
   b. Remove excess polish and rinse several times with acetone in an ultrasonic bath.
   c. Wear gloves while handling the shield to prevent contamination.

3. Reassemble the gun cartridge with a new filament.
   a. Carefully center the tip of the filament in the center of the shield aperture; this is usually accomplished with four set screws.
   b. Set the distance between the filament tip and shield according to manufacturer specifications.
   c. Using compressed air or Freon, remove all dust.
   d. Insert the gun cartridge into the electron gun.
4. Evaluate the system and align the column. Alignment is discussed later in this chapter; also consult the manufacturer's operating manual.

## Imaging System

The diameter of the electron beam as it exits the gun assembly is roughly 25,000-50,000 Å. This diameter must be narrowed to be an effective imaging probe. Accordingly, the electron beam is demagnified by a series of *convergent magnetic lenses* to a diameter of ~100 Å at the sample level (Fig. 1-2 and 1-3). In terms of current levels, the electron-gun current of $\sim 10^{-4}$ amp ($\sim 10^{15}$ e$^-$/sec) is reduced to $10^{-12}$ to $10^{-10}$ amp ($\sim 6 \times 10^{6}$ e$^-$/sec) at the sample level (Everhart and Hayes, 1972).

*Beam diameter,* also referred to as *spot size,* is one of the most important parameters for optimal SEM imaging, and it is controlled directly by the operator. As a rule, the smaller the spot size, the higher the resolution. The point-to-point resolution cannot exceed beam diameter: if a spot size of 500 Å is used for imaging, the maximum resolution attainable is 500 Å, which is quite poor. By decreasing the spot size to ~100 Å, both resolution (100 Å) and imaging are enhanced. Routinely, the spot size ranges from 100 to 200 Å, whereas high-resolution work requires that the spot size be decreased to ~50 Å. The only circumstance where it is advantageous to exceed a spot size of ~200 Å is when energy-dispersive spectroscopy is being conducted. As will be discussed, this enhances the intensity (count rate) of characteristic X-ray peaks.

An artifact frequently observed when the spot size is excessively large is *charging,* which is manifested as bright streaks or flashes across the width of the CRT or photograph. It results when the specimen is not connected to ground, i.e., the specimen accumulates a net negative charge. Charging is observed when nonconductive samples are examined at excessive accelerating voltage or spot size (Pawley, 1972; Pfefferkorn et al, 1972; Shaffner and Hearle, 1976); contaminant particles (e.g., airborne dust) or films may induce localized charging. This artifact is reduced or eliminated as follows:

1. Nonconductive samples may be coated with a thin film of conductive metal or carbon (see Chapter 8) or examined at low voltages.

2. The sample is connected to ground using a conductive adhesive (e.g., silver paint, colloidal carbon paint, or metal tape).
3. All samples should be cleaned with a blast of compressed air or Freon to dislodge adhering particles.
4. Inert samples (e.g., metals) should be cleaned with organic solvents to remove oils, grease, or any other soluble films (see Chapter 4).

Magnetic lenses are also responsible for focus and magnification of the image (see Siegel, 1975). *Focus* is achieved by varying the current passing through the final lens (traditionally referred to as the objective lens) and thus changing its focal length. The objective lens has a long focal length and is capable of imaging samples of various heights. There are also several levels of focus sensitivity available, from coarse to very fine focus. Images should always be focused at least two steps beyond the desired magnification level to ensure that the image recorded is truly in focus. It is much simpler to raise magnification, adjust the fine focus, and return to the desired level than to strain one's eyes trying to focus at only one level.

SEM *magnification* is the ratio of the size of the display area on the CRT to the distance the probe is scanned. Because a change in magnification simply involves scanning a different-size area, focus will be maintained when magnification is changed. Direct readouts of magnification are usually accompanied by micron markers from which dimensions may be measured. The micron markers are most useful when a micrograph is enlarged; the numerical magnification readout will obviously be incorrect if enlargements are prepared.

Either as part of or immediately beneath the final lens is a *deflection coil* which moves the electron beam across the specimen in a square or rectangular pattern. The beam is deflected by fields controlled by the *scan generator* and *deflection yoke* (Fig. 1-2), which are synchronized with the CRT. This synchronization results in a 1:1 correspondence between the position of the electron beam on the specimen and the image observed on the CRT screen. These time-sequencing events are largely responsible for the high depth of field observed in SEM micrographs.

The *scan speed*, or the rate at which the beam passes over the specimen, is variable from 100 to 100,000 lines/scan. Very rapid scan rates produce a static or nearly static image and are analogous to conventional television images. (Note that a standard American TV image is composed of 525 horizontal lines.) During slower scan rates the progression of the beam across the specimen is observed. Extremely slow scan rates are used to photograph the image. As a rule, slower scan rates improve image clarity because the electrons have sufficient time to interact with the specimen, which in turn releases more data signals.

Rapid scan rates are used for visual examination of the specimen to select regions of interest, focusing, column alignment, etc. — any imaging-related purpose except image recording. Moderate scan rates are used to evaluate focus and prepare for image recording: this visual rate produces an image which closely approximates the subsequent photograph. Slow scan rates (~30-120 sec/frame) are used to record the image; this topic is developed in Chapter 2.

*Astigmatism* is an optical aberration caused by minute flaws or inhomogeneities in the magnetic lens' coilings. Occurring when a lens focuses more strongly along one axis than along other axes, it is manifested as a distortion of shape as focus is varied. For example, a circle will become an ellipse on either side of focus if astigmatism has not been compensated for; therefore, astigmatism is formally defined as deviation from the theoretically perfect circular symmetry of the optical axis. The asymmetry is corrected by incorporating stigmators in the final lens. *Stigmators* are weak lenses which exert a magnetic field having a magnitude equal to but opposite from that of the asymmetric field generated by the final lens, thus canceling out the asymmetry caused by imperfections in the lens. Stigmators are of variable amplitude and direction, the effects of which are readily observed at magnifications exceeding 2000×. Because these are relatively weak lenses, astigmatism must be corrected at a minimum of 2000×, followed by a high-magnification (~10,000×) double-check. The method of correction is discussed below.

Immediately beneath the final lens is an *aperture* that intercepts electrons which are not part of the imaging beam. The aperture prevents stray electrons from striking the specimen, thereby reducing the level of background noise in the image. If an aperture were not included in the lens assembly, a problem referred to as spherical aberration would result. *Spherical aberration* arises because the fields set up by magnetic lenses are much stronger at their periphery than in their center. Consequently, electrons at the perimeter of the beam are more strongly influenced than electrons travelling within the center of the optical axis. Although the presence of an aperture intercepts these stray electrons and eliminates spherical aberration, apertures reduce theoretical resolution values by limiting the numerical aperture of the lens. In effect, the aperture limits the angle of the extreme rays that are allowed to pass.

Apertures are normally composed of platinum or molybdenum. They may be formed into disks or ribbons for, respectively, single- or multiple-hole apertures. Single-hole apertures are typically between 50 $\mu$m and 100 $\mu$m in diameter; multiple-aperture strips offer a range of sizes from ~50 $\mu$m to 300 $\mu$m in diameter. Resolution, depth of field, and image clarity are enhanced with smaller apertures. Larger apertures may be preferred when conducting an X-ray analysis, where a more intense probe is desirable.

In addition to the aperture beneath the final lens, many SEMs have several more apertures positioned along the length of the optical axis, all serving to intercept stray electrons. The apertures are usually held within the *column liner tube* extending from the anode to the bottom of the column. The liner prevents accumulation of contaminants on the lenses; contaminants arise during normal operation from residual gas and water vapor, sublimation of the sample during irradiation, vacuum pump oil, etc. Every three to four months the column liner tube and aperture must be cleaned or replaced for optimal instrument performance. If astigmatism cannot be compensated for by manipulation of the stigmators, the problem is not magnetic lens field asymmetry but contaminated apertures.

Molybdenum apertures are cleaned in a vacuum bell jar by heating to red hot, whereas platinum apertures are cleaned by submersion in fuming sulfuric acid. The procedure outlined below may be used for alignment; note that this same procedure may be used to evaluate alignment on a daily basis:

1. Saturate the filament and focus the image at ~500×. Observe image motion as focus is varied: if the image asymmetrically expands and contracts from center, adjust the $X$ and $Y$ movements of the final aperture to minimize that motion. Repeat at ~1000×.
2. Check the gun alignment for the brightest image.
3. At ~1000×, go through focus and again observe image displacement. If the image is asymmetrical, adjust the stigmator direction and amplitude to compensate for any residual astigmatism. Do not exceed the point of image symmetry.
4. Increase magnification to 5,000-10,000×, and repeat step 3.
5. If the desired magnification exceeds ~7,500×, increase magnification at least two steps beyond the desired level and repeat step 3.

To summarize, the imaging system of the SEM consists of a series of magnetic lenses and accessories which serve to:

1. Control beam diameter (spot size).
2. Focus and magnify the image.
3. Scan the beam in a regular fashion over the specimen.
4. Reduce spherical aberration with apertures.
5. Compensate for image asymmetry with stigmators.

## INFORMATION SYSTEM

The information system consists of (1) the sample, which releases a variety of data signals resulting from interaction with the imaging beam, and (2) a series of detectors which recognize and analyze the data signals.

The sample is mounted on a conductive substrate, usually an aluminum stub or carbon planchet, and then secured within the sample stage of the microscope. The stage serves as an electrical pathway to ground and is also equipped with several controls for specimen movement. The sample can be moved in the X, Y, and Z directions, tilted, and rotated. Obviously the X and Y axes (lateral motions) are manipulated to orient the specimen; the purposes of the other movements are not readily apparent. To appreciate specimen motion, one must first understand the origin of data signals and electron beam-specimen-detector geometry. Data signals result from interaction between the bombarding electrons and the atoms of the specimen (Fig. 1-5). Regardless of their format, data signals arise from either *elastic* (electron-nucleus) or *inelastic* (electron-electron) collisions of beam (primary) electrons with the atoms of a specimen. *Elastic collisions* will produce backscattered electrons (BSE), which provide both topographic and compositional information about the specimen (Newbury, 1977; Robinson and George, 1978). *Inelastic collisions* deposit energy within the sample, which then returns to the ground state by releasing distinct quanta of energy in the form of secondary electrons (2° e⁻ or SE), X-rays, light photons (cathodoluminescence), and such nonradiative transitions as phonon (heat) production. All of these events occur simultaneously. To avoid confusion, the origin and detection of each data signal will be discussed separately. An important factor to keep in mind is that although these signals are generated during irradiation, their analysis requires that appropriate

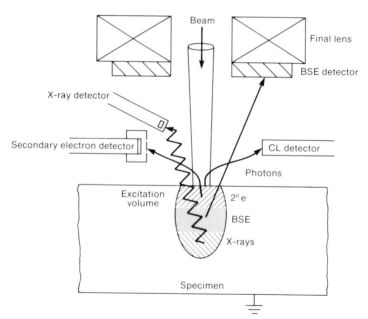

Fig. 1-5.   The origin and detection of data signals.

detectors be present; most microscopes cannot detect all of these data signals.

## Electron Signals

Inelastic (electron-electron) collisions between the beam electrons and the specimen electrons produce the *secondary electron imaging signal*. Essentially, the sample absorbs high-energy beam electrons and acquires a net negative charge. In order to reassume ground potential, the sample releases secondary electrons which are of significantly lower energy than the primary beam electrons. Multiple inelastic collisions result in the excitation of many specimen atoms to different levels of potential. The specimen accommodates these different excitation levels by releasing other forms of energy of lower potential than secondary electrons. Distinct quanta of energy in the form of X-rays and Auger electrons (both arising from inner-shell ionizations) and the re-combination of electron-hole pairs release light in the form of cathodo-luminescence: these are lower-energy transitions by which the specimen returns to ground.

In an elastic (electron-nucleus) collision, a primary electron strikes the nucleus of a specimen atom and rebounds with a negligible loss of energy (~20% below the incident beam energy; see Fig. 1-6) and a slight angular deflection (Rutherford scattering); because the same electron enters and exits the specimen, it is referred to as a *backscattered electron* (BSE). The trajectory of the BSE is analogous to the rebound of a tennis ball after it hits a cement surface: the ball bounces at a slight angle relative to the axis along which it proceeded downward. Further, if the tennis ball hits a grass surface rather than cement, it will not rebound to the same degree. This analogy also holds true in SEM: if the mean atomic weight (Z) of the specimen is low (e.g., plastics), the probability

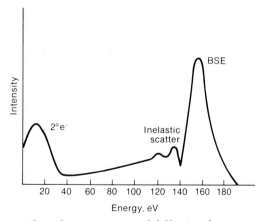

Fig. 1-6.  **The relative energies of different electrons.**

of a backscattering event is lower than if the specimen is of higher Z. Therefore, high-Z metals release a greater number of BSE than low-Z specimens.

The probability for an elastic scattering event is the *cross section σ*, which increases as the square of the atomic number and inversely as the square of the energy. The *backscattered coefficient, η*, defines the fraction of the beam electrons which escape from the specimen:

$$\eta = \frac{Z - 1.5}{6} \tag{1}$$

As shown in Fig. 1-7, $\eta$ increases as a function of Z and is independent of accelerating voltage in the 10-40 keV range.

*Backscattered electron imaging* provides a means of distinguishing zones of different atomic number within a composite specimen because each zone will exhibit a different contrast level: brighter areas are where a larger number of BSE are released and are thus of higher atomic number than darker (lower-Z) areas, where the emission level is reduced. Referred to as *atomic number imaging* or *atomic number contrast*, this is a powerful technique when used in conjunction with energy-dispersive X-ray analysis because visual as well as elemental information is revealed. Appropriate samples for BSE imaging include polished samples and samples in which there are multilayered metal platings in cross section. In the latter situation, it may be difficult to distinguish among the layers with conventional secondary electron imaging, whereas a BSE image carries different contrast levels for each plating level provided that the platings are of different atomic number. One may then define plating thickness, continuity, alloying, and composition through X-ray analysis of each layer.

Although we assume that SEM images arise strictly from surface phenomena, this is not exactly true: different information signals origi-

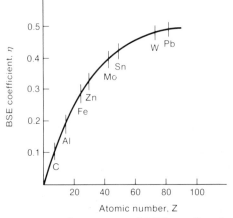

**Fig. 1-7.   The emission of BSE as a function of atomic number.**

nate from different depths in the sample, although the total excitation volume is commonly limited to the uppermost 100-200 $\mu$m of the specimen. The *excitation volume,* synonymous with *information depth* and *depth of penetration,* defines the volume where data signals originate and is a function of the atomic density of the specimen and accelerating voltage (Seiler, 1976). High-Z specimens prohibit deep penetration of the beam, and the excitation volume is hemispherical in shape (Fig. 1-8a). Low-Z specimens offer little resistance to the incident electrons, and the excitation volume assumes an elongated teardrop shape (Fig. 1-8b).

The excitation volume is subdivided into levels which correspond to the type of signal being emitted. Proceeding from the surface to the internal matrix, the following data signals are emitted: Auger electrons and cathodoluminescence, secondary electrons, backscattered electrons, and finally X-rays. Note that the BSE image originates from the zone closest to the X-ray volume: for this reason BSE images more closely approximate the origin of the X-rays than do conventional images.

The depth of penetration, $d_p$, is related to atomic number (Z) and accelerating voltage ($V_o$) by

$$d_p \propto \frac{W_a V_o^2}{Z \rho} \qquad (2)$$

where $W_a$ is atomic weight and $\rho$ is density.

As a rule, $d_p$ decreases as Z increases. Applying equation 2 to aluminum and tungsten examined at 30 keV, it is clear that the beam penetrates much further into lower-Z aluminum than into higher-Z tungsten:

$$\text{For Al:} \quad d_p = \frac{(26.98)(30)^2}{(13)(2.7)} = 691.86$$

$$\text{For W:} \quad d_p = \frac{(183.85)(30)^2}{(74)(19.3)} = 115.86$$

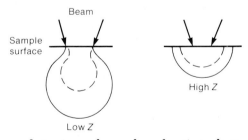

Fig. 1-8.   Influence of atomic number and accelerating voltage on the depth of the excitation volume. Low voltage, dashed line; high voltage, continuous line.

Another parameter influencing electron emission is the conductivity of the specimen. Metals are conductive and readily emit electron signals, but nonconductive specimens such as plastics, glasses, or ceramics do not behave in the same manner: the primary beam is absorbed by the sample, which accumulates a net negative charge of sufficient magnitude to deflect the primary beam, and consequently the image is poor. Beam absorption is suppressed by coating the sample surface with a conductive metal and providing a pathway to ground by means of conductive paint or tape. The preparation of conductive thin films (e.g., gold or carbon) by evaporation or sputter coating is the topic of another chapter. In addition, it is frequently desirable to reduce the accelerating voltage to moderate (10-20 keV) levels. When any coating is undesirable, adequate imaging may be obtained using very low (2-10 keV) accelerating voltages.

The secondary and backscattered electrons used for conventional imaging are detected by an *Everhart-Thornley (E-T) electron detector* (Everhart and Thornley, 1960). Virtually all SEMs employ this type of detector for imaging. The detector is at a 90° angle relative to the optical axis and is situated such that the specimen may be tilted in its direction. Tilting increases the number of electrons entering the detector and improves image quality.

A schematic of the Everhart-Thornley detector is shown in Fig. 1-9. Low-energy (~4 eV) secondary electrons are attracted toward the detector by a positively charged (40-200 V) *Faraday cage*. Secondary electrons therefore follow a curved trajectory toward the detector: this implicitly means that E-T detectors are not dependent upon line of sight. Backscattered electrons, which follow a line of sight, will enter only if their trajectory is within the solid angle of collection of the E-T detector.

Held within the Faraday cage is an aluminum *scintillator*. A potential difference of 10-12.5 keV accelerates the secondary electrons toward the scintillator, which converts the electrons into a proportional number of photons. The light signal propagates through a *light pipe* and to a *photomultiplier tube*, which amplifies and converts the light signal back into

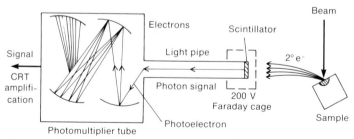

**Fig. 1-9.   Detection of electrons by the Everhart-Thornley detector.**

an electrical pulse. The electrical pulse (photocurrent signal) modulates the CRT screen brightness, and an image is produced.

The scintillator is basically an extremely thin layer of aluminum in contact with the end of the light pipe; as such, it is subject to a high contamination rate. A dirty scintillator reduces the efficiency of electron collection and degrades image clarity. Contamination originates from sublimation of low-Z specimens, evacuation of wet samples, and back-filling the specimen chamber with air rather than nitrogen. Even when contamination is not a problem, with normal use the scintillators will degrade. Optimal SEM imaging requires that the scintillator be replaced every three to four months (when apertures are also changed). Scintillators are expendable items available from SEM manufacturers and EM suppliers. Used scintillators may be cleaned and coated with a thin layer of evaporated aluminum; however, the film thickness is critical and consequently sophisticated thin-film measurement devices are required.

Backscattered electrons linearly rebound from the specimen and retain ~80% of the incident-beam energy, which eliminates the need for a Faraday cage. Everhart-Thornley detectors may be used for BSE imaging, but their orientation in the specimen chamber is not ideal. When this is the only electron detector available, BSE may be collected and secondaries rejected simply by reducing the potential of the Faraday cage to −50 V. Thus, one may easily compare BSE images to conventional images and learn more about a specimen.

The design of BSE detectors is based upon the cosine distribution of BSE around the primary beam (Fig. 1-10a). As such, BSE detectors follow a line of sight, with the proportion of BSE detected dependent upon the large solid angle subtended to the beam as well as proximity to the beam. Therefore, detectors designed solely for BSE detection typically are mounted beneath the final lens and surrounding the optical axis.

*Solid-state BSE detectors* (Fig. 1-10a) are composed of a wafer of doped semiconductor silicon across which are thin-film electrodes. When the semiconductor is struck by an electron, electron-hole pairs are created and swept apart by a bias before they can recombine. These in turn create a current directly proportional to the energy of the incident BSE, which is used as the image formation signal. Because the detector relies upon high-energy electron excitation, the low-energy secondary electrons are not detected.

The *Robinson detector* (Robinson, 1974), originally designed for the observation of wet biological samples, has found application as a BSE detector. The detector is a plastic scintillator designed such that it surrounds the sample surface, and is connected to a photomultiplier tube (PMT) for image formation (Fig. 1-10b). The Robinson detector is not fixed within the sample chamber, but movable toward or away from the

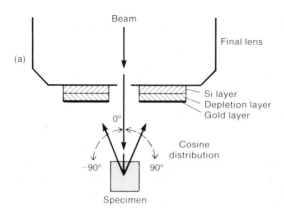

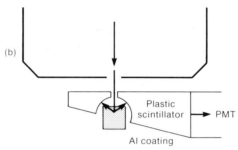

Fig. 1-10.   Detection of BSE using a solid-state detector (a) or the Robinson detector (b).

specimen. As such, it cannot simultaneously be used with other detectors (secondary electron or X-ray), which is a disadvantage. Nonetheless, it produces excellent BSE images.

Optimal SEM imaging depends heavily upon several geometrical considerations. Because backscattered electrons follow a line of sight, the specimens must be positioned in such a manner that the maximum amount of information is collected, i.e., that portions of the specimen are not shielded from the detector. In comparison, secondary electrons follow a curved trajectory, and it is desirable to tilt the specimen toward the Everhart-Thornley detector to enhance collection.

To obtain a BSE image with a solid-state or Robinson detector, the sample surface should be perpendicular to the beam and parallel to the detector. The specimen should also be close to the detector; manipulate the Z-axis to shorten the working distance. When examining very rough surfaces, it may be necessary to carefully manipulate the tilt or rotate movement to bring different surface features into view.

The beam-specimen-detector geometry is slightly more complicated when the Everhart-Thornley detector is used for imaging either backscattered or secondary electrons. Tilt is important because it influences the topographic contrast in the final image. For conventional imaging, the specimen is always tilted toward the detector because (1) more secondary electrons will be attracted toward the Faraday cage and enter the detector; (2) the detector is asymmetrical relative to the beam in that it is located at the side of the chamber; and (3) more backscattered electrons will be detected. Because the topography (i.e., smoothness or roughness) of different samples varies considerably, no rule exists for a standard angle, although it typically is between 15° and 45°. The novice should experiment with both rough- and smooth-surfaced samples and observe the effects of tilt on a given field of view at various magnification levels. Insufficient tilt lowers image clarity and contrast, while excessive tilt defocuses the upper and lower portions of a field of view.

Peaks, or those areas on the sample surface closest to the E-T detector, appear brighter than valleys or areas facing away from the detector. Deep crevices or pits in a surface may be difficult to image because the secondary electron signal is absorbed by the sample. (Note: As will be discussed in Chapter 2, gamma signal processing permits observation of pits and crevices.) Secondary electrons originating from non–line-of-sight positions are still detected because of the curved trajectory imposed by the Faraday cage. This is the reason why SEM images have large depth of field and appear three-dimensional. If the detector high voltage is made negative, however, a very different image will result. Only BSE will be detected, and BSE follow a line of sight. Figure 1-11 illustrates these effects. The high contrast of the BSE image is moderated when combined with the secondary electron signal (Volbert, 1982).

The effect of *working distance* (WD), defined as the distance separating the specimen surface from the final pole piece, also influences imaging. WD is modified by manipulating the Z-axis, which moves the sample upward and downward. A short WD (close to the final lens) decreases the depth of field, raises the lower limit of magnification, may reduce image clarity by interfering with secondary electron collection, and may limit specimen movement. On the other hand, a long working distance enhances the depth of field and permits very low magnification (~8×) observations, but it also reduces image clarity because the electron signals must travel farther for detection. Consequently, one compromises by using moderate working distances for imaging. At this position the specimen surface is either level with or slightly beneath the Everhart-Thornley detector. As will be discussed in Chapter 2, both working distance and angle of tilt are critical for an acceptable X-ray analysis, and what is suitable for imaging may not be appropriate for elemental analysis.

## X-ray and Cathodoluminescence Signals

A primary electron entering the sample matrix and undergoing in-elastic collisions can impart some of its energy to many atoms. As de-

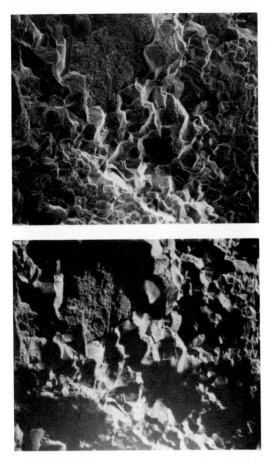

Fig. 1-11. Images recorded with the Everhart-Thornley detector. Top, combined secondary and backscattered electrons; bottom, backscattered electrons only. (Courtesy of A. Laudate and JEOL)

scribed above, a secondary electron may be ejected from an ionized atom, while lower-energy radiative transitions from atoms not sufficiently excited to eject an electron are *X-rays* and *cathodoluminescence* (light photons). Although light originates from the surface of opaque specimens, it may originate from anywhere within the excitation volume of a transparent specimen. X-rays originate from deep within the excitation volume (Fig. 1-5), and many are of sufficient energy to escape from the specimen and be detected. Because these are similar atomic transitions, they are discussed together; X-rays are dealt with more fully in Chapter 3.

During irradiation, a primary-beam electron may strike an atom, which ejects an inner-shell electron. To reassume ground potential, a higher-energy outer-shell electron falls into the vacancy and the atom simultaneously emits an X-ray having an energy equal to the energy difference between the two shells. Because that quantum of energy uniquely characterizes the atomic transitions of a given element, it is referred to as a *characteristic X-ray*. By measuring either the energy or the wavelength of the X-rays using, respectively, energy- or wavelength-dispersive spectrometers, identification of the element is possible.

The traditional *Bohr model* of the atom (Fig. 1-12) reveals that atomic structure includes a nucleus and at least one electron shell. *K-shell electrons* are closest to the nucleus and are ringed by *L-shell electrons*, which are in turn encircled by *M-shell electrons*. Because they are closest to the nucleus, K-shell electrons are more tightly bound than L- or M-shell electrons. The KLM shells are subdivided in $K\alpha$, $K\beta$, $L\alpha$, $L\beta$, etc., X-rays because not all electrons in a given shell possess exactly the same energy. A further subdivision into $K\alpha_1$, $K\alpha_2$, $L\alpha_1$, etc., accommodates smaller energy variations within a given shell. Iron releases characteristic X-rays having the following energies (keV): $L\alpha$, 0.704; $L\beta$, 0.717; $K\alpha_1$, 6.403; $K\alpha_2$, 6.390; and $K\beta$, 7.057. No other element will emit this exact energy sequence. $K\alpha$ X-rays are always the most intense signal and originate from an L-shell electron dropping into a vacant K shell (the K shell being vacant because of emission of a secondary electron). $K\beta$ X-rays are more energetic and arise from an M-shell to K-shell transition. $L\alpha$ X-rays are of lower energy and intensity and result from an M shell filling an L-shell vacancy; $L\beta$ X-rays result from N-shell to L-shell transitions.

The X-rays are detected by spectrometers which discriminate among either the energies or the wavelengths of the rays. Energy-dispersive spectrometers are most commonly associated with SEMs, while

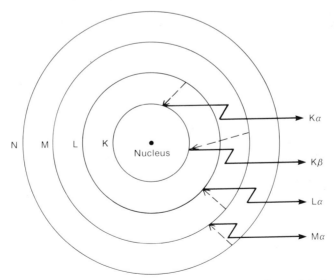

**Fig. 1-12.     The origin of X-rays as shown in the Bohr model of the atom.**

wavelength-dispersive spectrometers are coupled to microprobes. The importance of X-ray spectroscopy in SEM warrants much further discussion and is covered in a separate chapter.

*Cathodoluminescence* (CL), or light photon emission, is a low-energy radiative transition that excited atoms emit to reassume the ground state. While X-rays originate from deep within the excitation volume, CL is a surface phenomenon for opaque specimens (Fig. 1-5); in light-transparent specimens, photons may arise from anywhere within the excitation volume. Furthermore, the intensity of the CL signal is inversely proportional to atomic weight and the backscatter coefficient: as Z and $\eta$ increase, CL emission decreases. Thus, CL is a useful imaging mode for low-Z specimens such as silicon-based semiconductors. During irradiation, atoms in the semiconductor become excited and specimen electrons are raised from states in the filled valence band to the empty conduction band, thereby producing electron-hole pairs. While additional electron-hole pairs are generated during continued bombardment, some electron-hole pairs will recombine and release light photons. When the excitation is removed, recombination of electron-hole pairs renders the semiconductor passive.

Cathodoluminescence is a line-of-sight phenomenon that follows a cosine distribution around the primary-beam axis; the analogy to backscattered electrons is clear. Detectors consisting of *ellipsoidal lenses or mirrors* are positioned around the specimen to collect as much of the CL signal as possible (Fig. 1-13); again, note the similarity to the Robinson detector position. The signal is then conducted through a light pipe, and amplified for display or for spectral analysis. In the latter situation, the CL signal passes through interference filters, a monochromator, or an optical multichannel analyzer that measures the wavelength of the emitted light (Fig. 1-14). Unfortunately, the theoretical basis (*Moseley's Law*) for describing these radiative transitions predicts that CL emission will

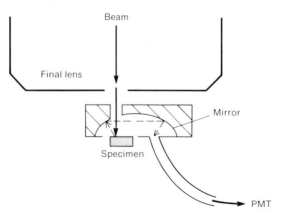

**Fig. 1-13.  Cathodoluminescence detection with an ellipsoidal mirror.**

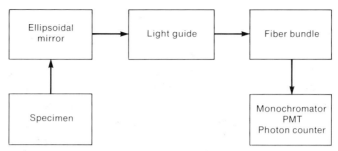

**Fig. 1-14.** Collection and analysis of cathodoluminescence.

cover a broad band (as opposed to X-ray peaks): consequently, quantitative measurements of CL spectra are theoretically improbable.

Because the cathodoluminescent mode is commonly used in semiconductor evaluation and not in the analysis of higher-Z materials, it will not be discussed further. Interested readers should consult Holt and Datta (1980) or Muir and Grant (1974) for further information.

## DISPLAY SYSTEM

SEM images are displayed on the screen of a *cathode-ray tube* (CRT), and permanent records or *scanning electron micrographs* are recorded by photographing either the visual CRT or a second high-resolution record CRT. Two CRTs are preferred because (1) *residual halation,* or the persistence of light emission by the fluorescent screen coating of the CRT after removal of excitation, will degrade the micrograph resolution by introducing noise; and (2) very fine-grained fluorescent coatings used in the record CRT screens afford finer resolution than more coarse-grained visual CRTs. The latter parameter has become slightly dated; current instruments having only one CRT have incorporated fine-grained, low-halation phosphors, thus eliminating these problems. However, in older instruments, it may be necessary to wait until the residual halation has disappeared before recording the image.

The synchronization between events occurring in the specimen chamber and those observed on the CRT screen were discussed under "Imaging System." Image recording is discussed in Chapter 2.

## VACUUM SYSTEM

The SEM optical column and specimen chamber are operated under high vacuum ($\geq 10^{-4}$ torr) for several reasons. First, residual gas molecules would scatter the electron beam, and the electrons would travel at different velocities, resulting in severe chromatic aberration. This in turn drastically limits image resolution. Operating under low vacuum would also result in accelerated oxidation of the tungsten filament, random electrical discharge along the optical axis, and contamination of the specimen. Consequently, the SEM is equipped with pumps that operate continuously to maintain high vacuum. The common high-vacuum systems used in SEMs are diffusion pumps and turbomolecular pumps, both of which are backed by low-vacuum rotary pumps.

## Rotary Pumps

The *rotary pump*, also known as a mechanical, rotary-vane, or roughing pump, is an oil-immersed, eccentric-vane pump which achieves low vacuum ($\sim 10^{-2}$ torr) by the simple displacement of air. Figure 1-15 is a cross section of the pumping chamber. The rotor contains spring-loaded vanes that maintain contact with the stator wall (housing) as the rotor revolves. Because the rotor is eccentrically mounted, a crescent-shaped gas volume is produced and the compressed gas is discharged during rotation. Rotation of the rotor produces a continuous cycle of suction, compression, and exhaust that results in the pumping action. In SEMs with diffusion pumps, the rotary pump has connections to both the diffusion pump and the microscope column, while in turbomolecular-pumped SEMs the rotary pump is usually attached only to the high-vacuum pump.

Some pump oil will escape during operation, and therefore the oil level of the rotary pump must be monitored and periodically replenished. Roughly once a year the oil should be drained and the pump filled with fresh oil. The fan belt should also be checked for signs of wear and replaced when necessary. Because rotary pumps are noisy and escaped oil vapor is a contaminant, they are frequently located outside of the SEM laboratory. The pump is then simply attached via vacuum hoses to the microscope.

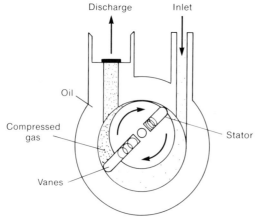

Fig. 1-15.   Cross section of the rotary pump.

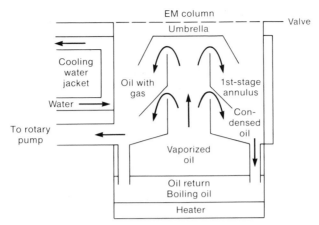

**Fig. 1-16.** Cross section of a high-vacuum diffusion pump.

## Diffusion Pumps

The *diffusion pump* is mounted directly beneath the microscope column and is backed by a rotary pump. A cross section of the pump is shown in Fig. 1-16. Liquid oil is heated at the base of the pump, vaporizes, and travels upward through a central tube. The vaporized oil stream strikes an *"umbrella"* and is deflected downward. The vapor-stream will entrap any gas molecules and carry them toward the base of the pump, where the gas is removed by the rotary pump. The vaporized oil strikes the cooled casing of the diffusion pump, condenses into a liquid, and gravitates toward the heater zone. The entire scenario is continuously repeated and a pressure differential, corresponding to high vacuum, is achieved.

Diffusion pumps normally have a series of umbrellas, the number corresponding to, for example, a three- or four-stage pump. Each stage is capable of producing a pressure difference of $10^{-1}$ torr; by arranging a three-stage diffusion pump in series with a rotary pump ($10^{-2}$ torr), the highest vacuum that theoretically can be achieved in a closed system is $10^{-5}$ torr.

The diffusion pump oil is a source of contamination for the SEM column, particularly in the event of a power or water failure. Most SEMs have interlocks which automatically seal the pump from the column if the pump overheats, but it is possible for oil vapor to diffuse into the column. Water recirculators may be used, particularly in areas where the tap water is aggressive, but a power failure obviously turns off both the SEM and the recirculator. There is no simple answer for this dilemma; simply be prepared for emergency shutdown of the SEM.

## Turbomolecular Pumps

*Turbomolecular pumps* are becoming popular as replacements for diffusion pumps on SEMs because they minimize contamination. As

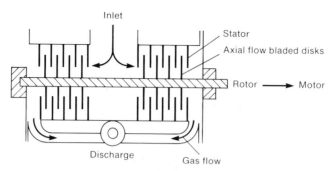

**Fig. 1-17.** Cross section of the turbomolecular pump.

such, the application of turbomolecular pumps in high-resolution SEM and quantitative EDS is significant. The pumping chamber resembles a turbine (Fig. 1-17) and consists of alternating rows of rotating blades and stationary slotted disks. When driven at high speed, the rotating blades will strike any gas molecules and propel the molecules toward the adjacent disk. The sequence continues through subsequent stages, and the gas is finally removed by the rotary pump. A very high pressure ratio is established by the alignment of the disks relative to one another, and the turbomolecular pump quickly and cleanly evacuates the microscope column.

The major reason why turbomolecular pumps have not completely replaced diffusion pumps is their high expense. Nevertheless, they are not a source of contamination, which is critical in some applications.

# SUMMARY: OPERATION AND MAINTENANCE OF THE SEM

The SEM can only perform as well as its operator. The operator must understand the operating principles of the SEM and maintain the system in optimal condition. The following brief discussion summarizes the previous pages and is applicable to all SEMs. Also refer to the manufacturer's operating manual for more specific instructions.

## SEM Operation

The basic operating conditions for imaging involve working under high vacuum with minimal contamination, saturating the filament to its effective operating temperature, and ensuring that the column is properly aligned. After achieving these conditions, the optimal imaging conditions for a given specimen are controlled by the accelerating voltage, spot size, and focus. Geometrical parameters such as working distance and tilt are also influential. Finally, the specimen must be prepared in such a manner that it is clean and conductive. Pfefferkorn et al (1978) presented an excellent overview of SEM operating parameters, many of which are included here.

Saturating the filament after high vacuum has been achieved ensures that the filament lifetime is maximum. If the filament is undersaturated, the column cannot be properly aligned, while at oversaturation the filament lifetime is reduced to a few hours. Although in most SEMs the filament cannot be heated until high vacuum is achieved, in older instruments the filament can be turned on below the most effective vacuum; this accelerates oxidation of the filament.

The accelerating voltage is chosen using the following guidelines: for examining metals, operate at 25-30 keV; nonconductive samples coated with a conductive thin film (e.g., gold sputtered replicas) are examined at 15-20 keV; and nonconductive specimens are examined at very low voltages (2-5 keV). Initially examine the specimen with a moderate spot size, increasing it for EDS or decreasing it for high-magnification work.

After the filament has been saturated, the column alignment is evaluated. Alignment should be checked for each and every specimen, and periodically during a prolonged examination. Gun alignment is evaluated in the reduced rapid or TV scan rate, or using a waveform monitor (if available). In either scan mode, manipulate the gun $X$ and $Y$ electronic controls until the image is uniformly bright and centered on the CRT. Using the waveform monitor, manipulate the gun controls until the wave is at its maximum height. If the gun $X$ and $Y$ electronic controls run out of movement, physically adjust the gun set screws (usually four) until the image peaks. The set screws are moved as opposing pairs, i.e., as one screw is moved outward the opposing screw is moved inward. The movement of the electronic controls is thereby re-established.

Next, in the rapid visual or TV mode, focus the image at about 1000×. Select a field of view where a symmetrical object (e.g., sphere or pit) is visible. Vary the focus and observe image motion: if the image does not symmetrically expand from its center, correct this asymmetry by manipulating the stigmators. Astigmatism has been corrected when the image uniformly expands and contracts as focus is varied. Microscopes having a "wobbler" control simplify this operation; simply switch the wobbler on before adjusting the stigmators. Again cancel out any asymmetry, then turn the wobbler off.

If asymmetry cannot be corrected by the stigmators, the final aperture is off-axis. Move the aperture $X$ and $Y$ controls while rotating through focus, then again correct the stigmators. It may be necessary to repeat this step several times before the image clarity improves. The same operations are conducted at ~10,000× and higher, if necessary, to ensure that the column is aligned over a broad magnification range. If the asymmetry cannot be corrected, the apertures must be replaced or cleaned.

Return to low magnification and generally evaluate the specimen's features. Adjust working distance and tilt to obtain the desired orientation and surface features. It is recommended that a low-magnification overview of the specimen be recorded, and this "map" be used to locate any subsequent photographs. While it may be difficult to interpret a single high-magnification photo, a series of photos from low to high magnification maintains perspective and orientation, thereby simplifying interpretation. This is particularly helpful when examining fracture surfaces.

When the examination is completed, the filament should be cooled for a minimum of 3 min before the chamber is brought to atmospheric pressure. Backfilling with nitrogen rather than air reduces contamination. The SEM is then evacuated and left in the high-vacuum condition. Completely turning off the instrument on a daily basis is not recommended; the SEM should be shut down only if it will not be used for several weeks, during certain maintenance procedures, or under emergency conditions.

## SEM Maintenance

During routine use, the filament is replaced roughly every 40 hours, while general column cleaning is necessary every three to four months. The methods for changing a filament and cleaning the gun cartridge were discussed on pages 9-10. The important factors in filament replacement are to clean the shield thoroughly, center the filament in the shield aperture, and set the distance between the filament tip and aperture according to manufacturer specifications.

The column liner tube, apertures, and detector scintillator become contaminated during routine SEM operation. Contamination arises predominantly from hydrocarbons originating in O-rings, gaskets, vacuum pumps, and lubricants. The vaporized hydrocarbons react with the electron beam and form carbonaceous deposits on the specimen and throughout the column. Another major source of contamination is the specimen, which may sublime during examination and outgas during evacuation. Clearly, the specimen and any adhesive (e.g., silver paint) must be dry before placement in the SEM. Similarly, do not handle the sample or any internal parts of the SEM unless you are wearing gloves; oils from one's hands are another source of contamination. The laboratory environment can contribute to contamination; this effect is minimized by backfilling the SEM with nitrogen and maintaining high vacuum at all times. Contamination is discussed in detail by Echlin (1975) and Miller (1978).

In those microscopes having a column liner, the liner is removed, polished, rinsed in several changes of high-grade acetone in an ultrasonic bath, and dried with compressed air or Freon. Dirty apertures are replaced as discussed on page 13. Finally the Everhart-Thornley detector's scintillator must be replaced. The latter procedure is conducted according to the manufacturer's operating manual.

The column is then reassembled, evacuated, and aligned as follows:

1. Using a resolution standard (e.g., gold test grid), saturate the filament and adjust the gun alignment controls.
2. Magnify the image to ~500× and evaluate and correct asymmetry by alternately manipulating the aperture X and Y movements and the stigmators.
3. Recheck gun alignment.
4. Repeat steps 2 and 3 at 10,000×.
5. Alignment is complete when the image symmetrically expands and contracts from its midpoint at a minimum of 10,000×.

# REFERENCES

Broers, A. N. (1975) Electron sources for scanning electron microscopy. *IITRI/SEM*, p 661.

Echlin, P. (1975) Contamination in the scanning electron microscope. *IITRI/SEM*, p 679.

Everhart, T. E., and T. L. Hayes (1972) The scanning electron microscope. *Sci. Amer.* 225:54.

Everhart, T. E., and R. E. M. Thornley (1960) Wide-band detector for micro-ampere low-energy electron currents. *J. Sci. Instrum.* 37:246.

Holt, D. B., and S. Datta (1980) The cathodoluminescent mode as an analytical technique: Its development and prospects. *SEM, Inc.* 1:259.

*Miller, D. E. (1978) SEM vacuum techniques and contamination measurement. *SEM, Inc.* 1:513.

Muir, M. D., and P. R. Grant (1974) Cathodoluminescence. In: *Quantitative Scanning Electron Microscopy*. (Holt, D. B., et al, eds.) Academic Press, London, p 287.

*Newbury, D. E. (1977) Fundamentals of scanning electron microscopy for physicist: Contrast mechanism. *IITRI/SEM* 1:553.

Pawley, J. B. (1972) Charging artifacts in the SEM. *IITRI/SEM*, p 153.

Pfefferkorn, G. F., et al (1972) Observations on the prevention of specimen charging. *IITRI/SEM*, p 147.

*_____ (1978) How to get the best from your SEM. *SEM, Inc.* 1:1.

Robinson, V. N. E. (1974) The construction and uses of an efficient back-scattered electron detector for SEM. *J. Phys. E.: Sci. Instrum.* 7:650.

Robinson, V. N. E., and E. P. George (1978) Electron scattering in the SEM. *SEM, Inc.* 1:859.

*Seiler, H. (1976) Determination of the "information depth" in the SEM. *IITRI/SEM* 1:9.

Shaffner, T. J., and J. W. S. Hearle (1976) Recent advances in understanding specimen charging. *IITRI/SEM* 1:61.

Siegel, B. (1975) Electron optics for scanning electron microscopy. *IITRI/SEM*, p 647.

*Volbert, B. (1982) Signal mixing techniques in scanning electron microscopy. *SEM, Inc.* 3:897.

---

*Recommended reading.

# 2

# Photography

Permanent SEM images are recorded by photographing the record CRT screen. The photograph, more correctly referred to as a *scanning electron micrograph*, reproduces the image observed on the visual CRT screen. Optimal recording presupposes that the user is correctly operating the SEM: photography cannot improve an inherently poor image. Black-and-white micrographs are the product of an SEM analysis, and the actual recording of these images is based upon conventional black-and-white photography. Photographic theory is well established and will not be presented here; interested readers should consult Horenstein (1974 and 1977), Neblette (1976), or other photography texts. This chapter will cover the practical aspects of SEM photography. A survey of signal processing, image processing, and stereo SEM is also presented.

## FILM TYPES

The most common recording medium used in SEM is Polaroid* film. Some microscopes are equipped with 35mm cameras which use conventional black-and-white film for image recording. The advantages of Polaroid film are that the photograph is developed immediately after exposure without additional processing; that either a positive print or a print and negative transparency are obtained; and, finally, that the negative is of high resolution. The major disadvantage of Polaroid film is that it is expensive. Nevertheless, a Polaroid camera back is standard equipment on virtually all SEMs.

The advantages of 35mm film are that slides for projection are easily recorded, a wide range of film speeds is available, and it is less expensive

---

*Copyright. Mention of a product name does not imply endorsement of that product.

than Polaroid film. The disadvantages of 35mm film are that roll processing is necessary and, if film is purchased in 24- or 36-frame rolls, the entire roll should be exposed before processing. Shorter lengths may be hand-rolled to avoid this problem. Scanning microscopists have thoroughly accepted Polaroid film as a recording medium, mainly to avoid the need for film processing and thereby save time and effort.

Two types of Polaroid film are available for SEM: *Type 52* produces a positive print and *Type 55* provides both a positive and a negative, the latter being used for enlargement and printing. Both film types are panchromatic, but they differ in speed and therefore in resolution. Type 52 film has an ASA of 400, DIN of 27, and resolution of 35-40 lines/mm. Type 55 film has an ASA of 50 and DIN of 18. The resolution of the print is 22-25 lines/mm and that of the negative 150-160 lines/mm. Because of its good resolution, the negative may be enlarged ($\sim 3\times$) and printed before "empty resolution" sets in.

For exposure to the visual CRT, one sheet of film (held in its protective packet) is gently slid into the camera back. Within the film packet is a sealed pod of developer; Polaroid cautions the user not to handle the film at the pod level. Once in the camera back, the light-tight envelope is withdrawn, the record CRT activated and the film exposed, the envelope reinserted, and the entire film pack withdrawn. The camera back has a control lever for opening and closing a set of rollers; when the control arm is in the "process" position, the rollers are in contact. As the exposed film is withdrawn, the rollers compress and break the developer pod, and the emulsion is evenly coated with a layer of developer. The film must be withdrawn with a continuous motion to ensure thorough coating; either jerking the film out or pulling too slowly adversely affects the film. Withdrawing the film too quickly usually leaves a portion of the photo undeveloped, whereas stopping during withdrawal leaves a vertical stripe across the micrograph. The presence of broad horizontal stripes on a photograph indicates that some developer has escaped from the film packet and leaked onto the camera rollers. When this occurs, remove the camera back, expose the rollers, and clean them with a soft cloth dampened with distilled water. Avoid contacting the developer because it is caustic and causes skin irritation; washing with water eliminates this problem.

The film develops within 20-25 sec after removal from the camera back, and the film pack is opened and the print (and negative, if Type 55 film is used) removed. To prevent image degradation, the print is evenly painted with coaters supplied with the film. The negative from Type 55 film must be treated as follows for preservation: after separation from the print, submerge the negative into aqueous 18% *sodium sulfite* and agitate for a few minutes. The negatives may be stored in the sodium sulfite bath for several hours, if necessary. Rinse the negative in running tap water (cool) for 5-10 min, then air dry. The negative is fragile, and care must be taken to avoid scratching the emulsion; commercially available film holders and processing tanks should be used. The dried negatives are stored in glassine envelopes to avoid damage.

## PRACTICAL SEM PHOTOGRAPHY

A good scanning electron micrograph will exhibit a range of gray tones, which are controlled by the contrast and brightness settings of the microscope. All microscopes have indicators for contrast and brightness, and the manufacturer usually recommends what levels to use for a good exposure. With experience, the microscopist visually determines the optimum exposure and ignores meter readings, which can vary. A few new instruments incorporate autocontrast and brightness systems.

The human eye can distinguish roughly 20-30 *gray levels,* which are gradations between black and white. Scanning electron microscopists are hampered, however, because residual halation of the visual CRT limits perception to about 12 gray levels. A broader gray scale can be reflected in the micrograph because the record CRT does not suffer from residual halation. Increasing the number of gray levels in an image enhances its information content. For this reason, SEM photos are normally recorded using the full width of the gray scale rather than extremes (black or white) of that scale. By this means the *signal-to-noise ratio* (SNR or S/R) is enhanced and image quality improved.

The SNR is also significantly controlled by the SEM operating parameters. A large portion of Chapter 1 was devoted to signal enhancement and noise reduction; recall that these factors are controlled by accelerating voltage, spot size, beam-specimen-detector geometry, aperture size, scan rate, and finally the specimen itself (cleanliness, conductivity, etc.). Obviously one attempts to enhance the image and suppress noise when recording that image: operating the SEM below the optimum will produce a poor micrograph. The following checklist serves to put image recording into perspective, and basically summarizes the operating parameters noted above:

1. An appropriate accelerating voltage should be selected based upon the mean atomic weight of the specimen: metals, 25-30 keV; nonconductive specimens coated with a conductive thin film, 10-20 keV; and nonconductive specimens, 2-5 keV.
2. At low and moderate magnifications, use a moderate spot size; as magnification increases, decrease the spot size.
3. The depth of focus is enhanced at low magnifications by using a small final aperture and large working distance; as magnification is increased, shorten the working distance to improve resolution.
4. Adjust specimen tilt to provide the maximum amount of data-signal collection.

5. The photographic scan rate should be adjusted according to the magnification level.
   a. Very slow scan rates ($\sim$120 sec/frame) increase the dwell time of the beam on the specimen surface and therefore enhance the signal-to-noise ratio for high-magnification ($\geq$10,000$\times$) recordings.
   b. Moderate scan rates ($\sim$60-80 sec/frame) are used for recording intermediate magnification levels (1,000-10,000$\times$).
   c. Rapid scan rates ($\sim$30-60 sec/frame) are used for recording low-magnification images.
6. A series of micrographs ranging from low to high magnification provides much more information than does a single high-magnification photograph. Perspective and orientation are retained, which are extremely important with certain types of specimens (e.g., fracture surfaces).

At this point the SEM novice should conduct a simple experiment to evaluate the effects of spot size and scan rate on a given specimen. In the TV or very rapid visual mode, observe the image at $\sim$500$\times$, first with a large and then a gradually smaller spot size. Adjust contrast and brightness as necessary, and record the large and small spot-size images. It will be observed that a large spot size provides excellent TV images but poor micrographs, and a small spot size degrades the TV image but enhances the micrograph. It can clearly be seen that spot size and scan rate are interrelated; successful scanning microscopists quickly learn to balance these factors. Table 2-1 lists some of the common problems in SEM image recording, giving causes and remedies.

## IMAGE ENHANCEMENT

Described below are two sets of techniques which serve to improve the data produced by the SEM. Signal processing refers to various modulations of the electron signal before it enters the SEM display system; i.e., the signal is modified within the microscope. Image-processing techniques enhance image clarity or quantify data within the image independent of the microscope.

### Signal Processing

SEMs usually have at least one signal-processing device which electronically manipulates the electron signal, and the modulated image is displayed on the visual CRT. Signal processing improves the clarity of specimens having excessively rough or excessively smooth surfaces: while an acceptable image may be recorded simply by adjusting contrast and brightness, one may enhance certain features (while suppressing other features) and improve the signal-to-noise ratio.

**Table 2-1.    Problems and Their Solutions in SEM Photography**

| Problem | Cause | Solution |
|---|---|---|
| Bright horizontal stripes | Charging of nonconductive material | Decrease spot size; clean specimen surface; coat nonconductive samples |
| | Poor contact with ground | Use conductive adhesive |
| | Poor vacuum | Dry specimen before evacuating; check vacuum seals |
| Horizontal bands | Dirty film rollers | Clean rollers |
| Vertical streak | Discontinuous film removal from camera back | Remove film pack with one steady motion |
| Poor resolution | Unsaturated filament | Saturate filament |
| | Large spot size | Decrease spot size |
| | Misaligned column | Align electron optical axis |
| | Astigmatism | Adjust apertures and stigmators |
| | Contaminated column | Clean column liner and replace scintillator and apertures |
| | Poor focus | Magnify 2 steps above desired magnification and focus |
| | Poor specimen geometry | Increase tilt and decrease working distance |
| Poor depth of focus | Aperture diameter too large | Use small final apertures |
| | Working distance too small | Increase working distance |

*Gamma modulation* is a common signal-processing device, applied when details of a specimen appear washed out because of high brightness (e.g., crests or the edge of the specimen) or when features are lost in darkness (e.g., the structure of pits). In other words, the intermediate gray levels have been lost. Gamma modulation suppresses very dark or very light levels and intensifies the intermediate gray levels. This effect is produced by conversion of the normal linear input signal into a logarithmic function which compresses the overall signal level (Fig. 2-1a). A gamma of 1 corresponds to the normal signal; below 1 the contrast gain is half as strong as that of brightness, and dark regions are amplified. With an increase in gamma, the contrast gain is twice that of brightness, and bright regions are amplified (Fig. 2-1b). Therefore, gamma <1 is useful for studying rough-surfaced specimens, while gamma >1 enhances the topography of smooth surfaces (Fig. 2-2).

Another mode of signal processing, *black-level subtraction*, relies upon differential amplification to enhance the number of gray levels in an image. Basically, a fixed dc component is subtracted from the signal,

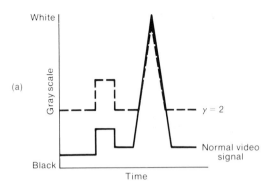

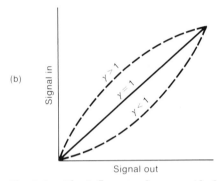

**Fig. 2-1.** The influence of gamma (dashed line) on the normal video signal (continuous line). (a) Compression of the video signal by a gamma of 2. (b) Log conversion of the linear video signal.

which is then linearly amplified and expanded across a larger number of gray levels (Fig. 2-3). Black-level subtraction will enhance image clarity, but excessive subtraction will degrade resolution by introducing excessive graininess into the image. This mode is most often used with backscattered electron images that exhibit poor contrast (Fig. 2-4).

Neither gamma modulation nor black-level subtraction drastically changes the appearance of a specimen surface. *Y-modulation,* on the other hand, is a mode of signal processing that produces a significantly different image. During normal image formation, the contrast observed on the CRT screen arises from beam-specimen effects. When Y-modulation is applied, the image is displayed as a set of closely spaced lines that essentially correspond to an intensity contour map of the surface (Fig. 2-5). Y-modulation therefore enhances surface texture and is useful for examining polished or smooth specimens. Y-modulation distinguishes only between low and high spots in the topography; because some distortion is introduced, such images must not be used for spatial measurements.

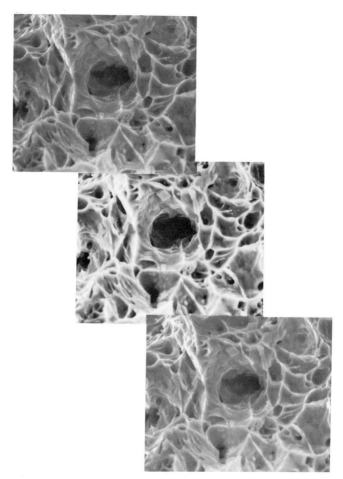

Fig. 2-2.   Micrographs showing the influence of gamma on the video signal. Top, amplification of dark regions with decrease in gamma; center, no gamma; bottom, amplification of brightness with increase in gamma. (Courtesy of A. Laudate and JEOL)

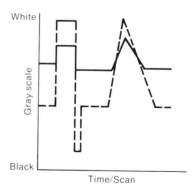

Fig. 2-3.   Black-level subtraction (dashed line) of the normal video signal (continuous line).

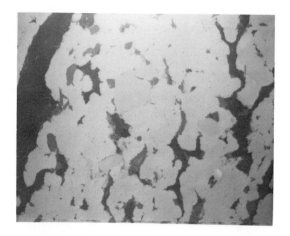

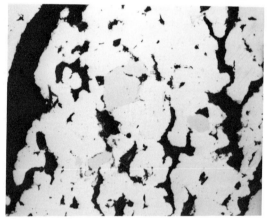

Fig. 2-4.   The effect of black-level subtraction on a BSE image
of a polished sample. Top, BSE image; bottom, subtracted BSE
image. (Courtesy of A. Laudate and JEOL)

*Signal inversion* is used to reverse black and white, so that negative
rather than positive images are recorded. This mode is useful for preparing transparencies for projection and also for examining negative
replicas, which are discussed in Chapter 7.

Other electronic signal-processing modes include *dual magnification*
or image display and *scan rotation* devices. Dual magnification may be
used to orient the location of a high- to low-magnification photo. A dual
image display also permits side-by-side imaging of, for example, a conventional electron image with its corresponding BSE image.

Scan rotation is an electronic movement of the beam that supplants
specimen-stage movement. For example, if an interesting surface feature is partially hidden, it may be awkward to move the specimen at
high-magnification levels without losing that feature. By activating scan
rotation, the given field of view may be reoriented without the risk
of displacement.

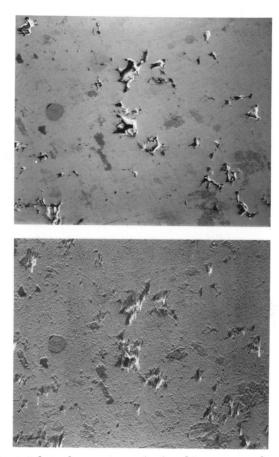

Fig. 2-5. A secondary electron image (top) and its corresponding Y-modulated image. (Courtesy of A. Laudate and JEOL)

These and other methods of signal processing are discussed in detail by Fiori et al (1974), Yew and Pease (1974), Newbury (1975), and Derksen (1981). The interested reader should consult these references for several imaginative applications, and also experiment with the different modes until they become familiar.

## Image Processing

Image processing can be subdivided into analytical image processing, which serves to extract quantitative dimensional information from an image, and digital image processing, which serves to enhance image quality. These techniques have applications in everything from microscopy to aerial photography (Attle et al, 1980; Cannon and Hunt, 1981). Probably the most unlikely application of image processing is for quality control of cookie size and the number of contained chocolate chips used by a major food supply company. The diverse applications of image processing have stimulated the development of instrumentation, which in turn has reduced its cost, making image processing a practical reality in service-oriented SEM facilities (e.g., Russ and Stewart, 1983).

As its name implies, the purpose of *analytical image processing* is to analyze and extract quantitative data from an image. The data typically are dimensional measurements such as area, perimeter, and aspect ratio measurements (e.g., Attle et al, 1980). This implies that even though an image conveys a significant amount of qualitative information, only a portion of it is quantitatively meaningful: analytical image processing serves to separate meaningful from irrelevant data.

An analytical image processor may be hard-wired (dedicated) or a general-purpose, nondedicated system (Jones and Smith, 1978). Dedicated systems are capable of analyzing real-time or stored images, whereas nondedicated systems analyze stored data (e.g., a micrograph or magnetic tape recording). Regardless of the format, the analyzer first isolates the desired feature from background. This operation is performed either by automatic gray-level distinction as interpreted by the instrument, or by the operator's perception of the image and manual entry of data points. Automatic processing is valuable in grain sizing, where the contrast levels of grain boundaries are very different (particularly in backscattered electron images). Here, the analyzer subdivides the image into smaller segments (pixels) which are distinct from background, compares the gray level corresponding to the desired feature, and converts the data into binary form. Digitization into a 256-by-256 point array defines the desired range of gray levels, and the data are quantitatively reduced into the desired format. For example, in a grain-size analysis, the analyzer would detect the grain boundaries, remove the irrelevant information, calculate the area of each grain per unit area, and display the quantitative data (Slater and Ralph, 1976; Gregory, 1983). By correlating this microstructural data with the materials properties, a very complete description of a specimen is possible. Applications such as this have made automatic processing an invaluable tool for metallography and quality control.

Manual data entry is preferred when the desired feature is not fully separable from background. For example, in a secondary electron image of a particulate specimen, a given particle may exhibit a broad gray scale that encompasses the background gray level: human perception can readily define the perimeter of the particle, whereas the analyzer cannot validly distinguish the entire area of the particle from the background. Manual entry is performed by displaying the image on a special graphics tablet sensitive to the pressure of a stylus which the operator uses to trace the outline of the desired feature. The analyzer then interprets and reduces these image points to quantitative data (Fig. 2-6). Powders, particles, and fibers are readily analyzed using this method.

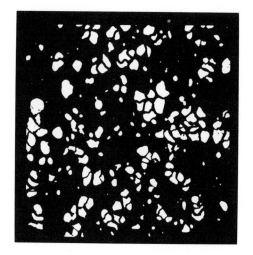

(a)

(b)

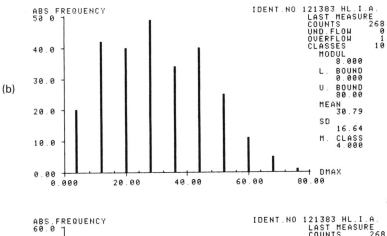

(c)

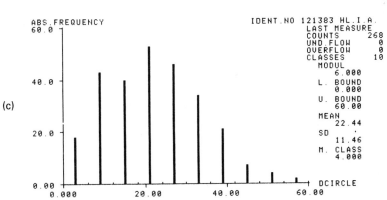

Fig. 2-6.    Image analysis of a metallic powder. (a) Photograph of the sampling showing the particles after digital enhancement. (b) Distribution histogram of maximum diameter (in nanometers). (c) Distribution histogram of equivalent circle diameter. (Courtesy of H. Loeb of Carl Zeiss, Inc.)

*Digital image processing* is a method of enhancement used to improve the visual appearance of an image (Jones and Smith, 1978). Until recently, image enhancement was limited to research situations, such as improving the clarity of satellite photographs or STEM images at atomic-level resolution (Crewe et al, 1980). The advent of readily available, inexpensive microcomputers has reversed this situation. Currently, several digital image processors have become commercially available at moderate cost; these improve image clarity beyond that afforded by signal-processing devices and conventional photographic techniques (e.g., Hardy et al, 1982). The manipulation of gray scale is one such operation; recall that degradations of the image by instrumental factors and limitations of film restrict the number of scale levels that can be displayed. Gray scales can be expanded several thousandfold by digitally recording a given image several times (in place of a single micrograph), and manipulating the range of gray scales until an image having the optimal signal-to-noise ratio is displayed. Using this method, structural details which are invisible in a conventional micrograph are enhanced. Note that such details may be recorded within a conventional micrograph but that the summation of degradations conceals them: digital image processing cancels out noisy manifestations. Theoretical discussions of digital image processing are beyond the scope of this text; interested readers should consult Gonzales and Wintz (1977), Hawkes (1980), IEEE (1981), and Kanaya et al (1982a, 1982b). Those desiring an overview of image processing in many applications will find Cannon and Hunt (1981) interesting.

## STEREO SEM

Although SEM images appear three-dimensional, their very format reduces them to two-dimensional representations. Their multi-dimensional appearance is due to high depth of focus, but perspective distortions introduced by the geometry of the beam/specimen/detector invalidate spatial measurements (both height and lateral dimensions). Our subjective impressions are based on the direction of illumination, which cannot be adequately described in a single micrograph produced by a complex geometry. These problems are aggravated when rough-surfaced specimens are examined, because the exact angle of a field of view is both unknown and unmeasurable. This in turn implies that magnification varies within a field of view.

The phenomenon of perspective distortion may be observed by recording an image having two prominent features at 20°, recording a second micrograph of the same field at 45°, and then measuring the distances between the two features on each micrograph: the measurements will differ and neither is valid. *Stereo imaging* involves recording a given field of view twice at slightly different orientations, and simultaneously viewing the stereo pair such that a three-dimensional image is perceived. Perspective is restored, and valid spatial judgments or measurements replace subjective impressions.

## Recording and Viewing Stereo Images

The four methods used to record stereo pairs are (1) the tilt method, where an angle is applied between the two micrographs; (2) the lateral-shift method, where there is a horizontal displacement between the two micrographs; (3) the rotation method, where a specimen is rotated between exposures; and (4) electromagnetic deflection of the electron beam between images (Boyde, 1975; Chatfield, 1978; Wergin and Pawley, 1980). Methods 1 and 2 are readily applied in any SEM, method 3 is difficult, and method 4 requires special accessories for the SEM; discussed below are the tilt and lateral-shift methods of stereo image recording.

The *tilt method* of stereo recording is a versatile technique that may be used in any type of SEM. It is desirable but not required to have a eucentric-tilt specimen stage (the tilt axis passes through the center of the specimen, not through the center of the stage). The tilt method is as follows: select and record the desired field of view, noting the tilt value of the specimen stage. With a wax pencil, mark the location of a prominent surface feature on the observation screen. Tilt the specimen approximately 7° (the stereo angle; see Fig. 2-7 and 2-8) while manipulating the stage $X$ and $Y$ axes to maintain the same field of view. Align the prominent surface feature beneath the wax pencil mark. Refocus the image using the Z-axis control; do not refocus with the objective lens controls. Adjust the contrast and brightness levels to match the first micrograph, and record the image.

The choice of the *stereo angle,* or the tilt difference between each micrograph of the stereo pair, is a function of the topography of the specimen. In general, smooth specimens require a stereo angle of 7-15°, while rough specimens require a stereo angle of 3-7°. The stereo angle determines *parallax* (synonym: horizontal displacement), which is a measure of the vertical position above or below a particular datum plane. With too large a stereo angle there is excessive displacement between features (i.e., excessive parallax), and the stereo image has "too much depth." With too small a stereo angle there is insufficient displacement, and the stereo image is not a true three-dimensional representation. If the microscopist is unsure of the optimal stereo angle for a given field of view, simply record several micrographs while changing the angle until the desired stereo image is obtained.

Another method useful for recording stereo pairs below 50× magnification is the *lateral-shift method* (synonyms: linear displacement or shift; translational shift). In this method, a micrograph is recorded, and the image is then moved horizontally while keeping the desired feature

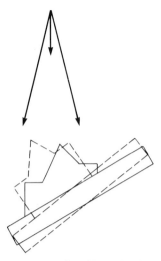

Fig. 2-7.    The effect of a 7° stereo angle. Continuous line, specimen at 30°; dashed line, specimen tilted to 37°.

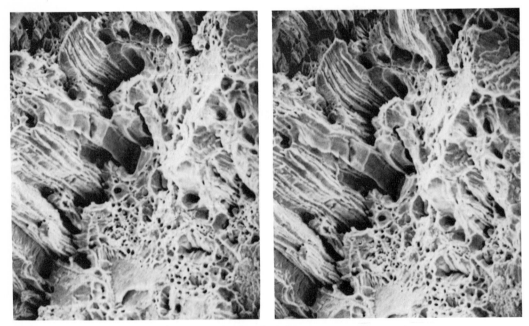

Fig. 2-8.    A stereo pair of a ductile fracture prepared using the
tilt method and a stereo angle of 7°. 300×.

in the field of view. The second member of the stereo pair is then
recorded. The distance shifted determines the depth of the stereo image;
a very large lateral shift produces more depth than a small shift. Operating conditions must be the same when recording each half of the stereo
pair; again, refocus (if necessary) by manipulating the Z axis, adjust
contrast and brightness to match the first micrograph, and maintain the

Fig. 2-9.   A stereo pair of a fractured wire prepared with the
lateral-shift method. 75×.

same tilt angle for both recordings. This method is appropriate only for
low-magnification images, because at moderate or high levels lateral
shift may displace the desired field of view before the optimal stereo
image is visible (Fig. 2-9).

Stereo pairs are viewed using simple pocket viewers, double-prism
viewers, or a mirror stereoscope (Boyde, 1979). Pocket viewers are
adequate for simple viewing of stereo pairs, but the more sophisticated
prism and mirror stereoscopes offer significant advantages when stereo
analysis is routinely conducted. The stereo effect (stereopsis) is perceived by positioning both micrographs within the viewer such that
the tilt axis is vertical; i.e., rotate both micrographs 90°. The micrograph with the lower tilt value is placed to the left of the second half of
the pair, and the distance between them adjusted until the stereo image
is prominent.

Because stereo viewers are designed for individual use, different
methods have been developed to simultaneously project stereo pairs.
The two most common projection techniques are the polarized and
anaglyph methods. The *polarized method* of stereo projection involves
the simultaneous projection of each member of a stereo pair through
adjacent slide projectors onto a lenticular silver screen. The projectors
are equipped with filters that polarize the image from one projector at
45° to the vertical and from the other at an angle perpendicular to this.
The audience must wear similarly polarized lenses to perceive the stereo
effect. Wergin and Pawley (1980) thoroughly discuss the methods and
equipment of the polarized method.

The *anaglyph method* of stereo projection applies a different color
(usually red or green) to each half of the stereo pair. The simultaneous
projection of the color-coded images produces a stereo image that is
perceived by individuals wearing red-green lenses. This was the method

used to project the 1950's 3-D horror movies. This method is not as popular as the polarized method, because color breakthrough and other problems may degrade the stereo effect (Barber and Brett, 1982). Nemanic (1974) and Barber and Emerson (1980) review the preparation and presentation of anaglyphs.

## Quantitative Stereoscopy

In addition to restoring perspective, stereo images may be used for spatial measurements. Referred to as *photogrammetry* or *quantitative stereoscopy,* valid measurements derived from stereo pairs are used in the construction of three-dimensional models. Boyde and his colleagues have contributed extensively to SEM stereoscopy (Howell and Boyde, 1972; Boyde, 1974, 1981; Howell, 1975; and other references cited throughout this chapter); summarized below are the more basic principles of quantitative stereoscopy.

The first stage of photogrammetry is establishment of the location of the *principle point,* defined as the point where the central ray (principle projector) of the scanning raster intersects the photographic plane. The location of the principle point does not necessarily correspond to the center of the micrograph; its location may be defined by switching off the SEM scan coils. The position of the stationary beam is visible as a bright spot. Alternatively, the position of the principle point can be established on a specimen by intentionally carbon-contaminating the surface: increase magnification roughly ten times above the desired level, permit a brief dwell time, and reduce the magnification to the desired level. A darkened spot (the site of contamination) is visible on beam-sensitive specimens, and corresponds to the location of the principle point.

Next, a *datum plane* which can be used as a baseline for measurements is identified with the tilt method; the common conventions use a datum plane parallel to one or the other halves of the stereo pair, or the midplane between the two micrographs. Using the parallel datum plane of one micrograph (Fig. 2-10), Howell and Boyde (1972) defined the following spatial measurements, where $M$ is magnification:

$$MZ_L = \frac{(MD)^2 [x_L \cos \alpha - x_R] - MD x_L x_R \sin \alpha}{[(MD)^2 = x_L x_R] \sin \alpha + MD(x_L - x_R) \cos \alpha}$$

$$Mx_L = \frac{x_L (MD - Z_L)}{MD}$$

$$MY_L = \frac{y_L (MD - Z_L)}{MD}$$

Figure 2-10 also shows the convention where an imaginary datum midplane between the stereo pair is located, and the following coordinates are shown:

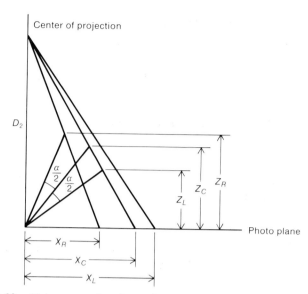

**Fig. 2-10.** Trigonometric relationship of the Howell and Boyde method. (Courtesy of Howell and Boyde, 1972)

$$MZ_c = \frac{(MD)^2(x_L - x_R)\cos\dfrac{\alpha}{2} + 2MDx_L x_R \sin\dfrac{\alpha}{2}}{[(MD)^2 + x_L x_R]\sin\alpha + MD(x_L - x_R)\cos\alpha}$$

$$MX_c = \frac{(MD)^2(x_L + x_R)\sin\dfrac{\alpha}{2}}{[(MD)^2 + x_L x_R]\sin\alpha + MD(x_L - x_R)\cos\alpha}$$

$$MY_c = \frac{Y_L}{MD}\left[MD + X_c \sin\frac{\alpha}{2} - Z_c \cos\frac{\alpha}{2}\right]$$

Individuals who regularly perform quantitative SEM measurements will be pleased to learn that modestly priced microcomputer programs are available for this purpose. Various systems are described by Howell and Boyde (1980), Roberts and Page (1980), Howell (1981), Boyde (1981), and Russ and Stewart (1983). Future developments in stereo SEM include real-time stereo imaging and image recording, computer control of the specimen stage, and continued acceptance and utilization by microscopists.

# REFERENCES

Attle, J. R., et al (1980) Applications of image analysis. *Amer. Lab.* 13(4):85.

Barber, V. C., and D. A. L. Brett (1982) "Colour bombardment"—a human visual problem that interferes with the viewing of anaglyph stereo materials. *SEM, Inc.* 2:495.

Barber, V. C., and C. J. Emerson (1980) Preparation of SEM anaglyph stereo material for use in teaching and research. *Scanning* 3:202.

Boyde, A. (1974) Photogrammetry of stereopair SEM images using separate images from the two images. *IITRI/SEM*, p 101.

⎯⎯⎯ (1975) Measurement of specimen height difference and beam tilt angle in anaglyph real time stereo TV SEM system. *IITRI/SEM*, p 189.

*⎯⎯⎯ (1979) The perception and measurement of depth in the SEM. *SEM, Inc.* 2:67.

⎯⎯⎯ (1981) Recent developments in stereo SEM (1981 update). *SEM, Inc.* 1:91.

Cannon, T. M., and B. R. Hunt (1981) Image processing by computer. *Sci. Amer.* 245(4):214.

Chatfield, E. J. (1978) Introduction to stereo scanning electron microscopy. In: *Principles and Techniques of Scanning Electron Microscopy*, vol 6. (Hayat, M. A., ed.) Van Nostrand Reinhold, New York, p 47.

Crewe, A. V., et al (1980) Further uses of color in scanning transmission electron microscopy. *Proc. 38th EMSA*, p 48.

Derksen, W. (1981) Signal processing devices for the SEM. In: *Microscopy/ Environmental Analysis*, ser 3, vol 3. International Scientific Communications, Inc., Fairfield, CT, p 154.

Fiori, C. E., et al (1974) Some techniques of signal processing in scanning electron microscopy. *IITRI/SEM*, p 167.

Gonzales, R. C., and P. Wintz (1977) *Digital Image Processing*. Addison-Wesley, Reading, PA.

Gregory, P. (1983) Advances in automatic image analysis. *Amer. Lab.* 15(4):29.

*Hardy, W., et al (1982) Digital image processing: A path to better pictures. *SEM,Inc.* 2:485.

Hawkes, P. A. (1980) *Computer Processing of Electron Microscope Images*. Springer-Verlag, New York.

Horenstein, H. (1974) *Black and White Photography: A Basic Manual*. Little, Brown and Co., Boston.

⎯⎯⎯ (1977) *Beyond Basic Photography*. Little, Brown and Co., Boston.

Howell, P. G. T. (1975) Taking, presenting, and treating stereo data from the SEM. *IITRI/SEM*, p 697.

⎯⎯⎯ (1981) Semi-automatic profiling from SEM stereopairs. *Scanning* 4:40.

*Howell, P. G. T., and A. Boyde (1972) Comparison of various methods for reducing measurements from stereo-pair scanning electron micrographs to "real 3-D data." *IITRI/SEM*, p 233.

⎯⎯⎯ (1980) The use of an XY digitiser in SEM photogrammetry. *Scanning* 3:218.

IEEE (1981) Special issue on image processing. *Proc. IEEE* 69(5).

*Jones, A. V., and K. C. A. Smith (1978) Image processing for scanning microscopists. *SEM, Inc.* 1:13.

Kanaya, K., et al (1982a) Digital processing methods using scanning densitometer and microcomputer for the structural analysis of a scanning electron micrograph. *SEM, Inc.* 1:61.

⎯⎯⎯ (1982b) Digital processing methods for structural analysis of an electron micrograph. *SEM, Inc.* 4:1395.

Neblette, C. B., et al (1976) *Photography: Its Materials and Processes*. Van Nostrand Reinhold, New York.

Nemanic, M. K. (1974) Preparation of stereo slides from electron micrograph stereopairs. In: *Principles and Techniques of Scanning Electron Microscopy*, vol 1. (Hayat, M.A., ed.) Van Nostrand Reinhold, New York, p 135.

Newbury, D. E. (1975) Techniques of signal processing in the scanning electron microscope. *IITRI/SEM*, p 727.

Roberts, S. G., and T. F. Page (1980) A microcomputer-based system for stereo-grammetric analysis. *Proc. Roy. Micros. Soc.* (Micro 80 suppl.) 15(5):8.

*Russ, J. C., and W. D. Stewart (1983) Quantitative image measurement using a microcomputer system. *Amer. Lab.* 15(12):70.

Slater, J., and B. Ralph (1976) The status of automatic image analysis in materials science and technology. *Amer. Lab.* 8(12):41.

*Wergin, W. P., and J. B. Pawley (1980) Recording and projecting stereo pairs of scanning electron micrographs. *SEM, Inc.* 1:239.

Yew, N. C., and D. E. Pease (1974) Signal storage and enhancement techniques for the SEM. *IITRI/SEM*, p 191.

---

*Recommended reading.

# =3=

# Energy-Dispersive Spectroscopy

The SEM is frequently equipped with a spectrometer capable of detecting X-rays emitted by the specimen during electron-beam excitation. These X-rays carry a characteristic energy and wavelength, which when measured will reveal the elemental composition of the specimen. The ability of the SEM to combine surface morphology information with X-ray microanalysis is unique. Unlike other electron optical microanalytical techniques (e.g., electron diffraction of microcrystals), X-ray analysis can be performed on a wide variety of sample types containing elements that encompass a large portion of the periodic table. Further, the minimum detectability limit of X-ray analysis is roughly $10^{-16}$ gm. Limitations of X-ray analysis include the inability to analyze light elements (e.g., C, N, O) under typical laboratory conditions, thereby eliminating the analysis of many organic compounds; and the inability to specifically identify compounds or the ionic state of the detected element. For example, one may analyze rust and feel comfortable referring to an iron-base oxide, but the distinction between the ferric and ferrous states cannot be made.

The two basic types of X-ray microanalysis used in conjunction with SEMs are *energy-dispersive spectroscopy* (EDS; also referred to as energy-dispersive X-ray analysis, EDXRA) and *wavelength-dispersive spectroscopy* (WDS), which discriminate among the energies or wavelengths of X-rays, respectively. Energy-dispersive systems are more commonly associated with SEM than are wavelength-dispersive systems, although the latter are becoming increasingly popular (WDS is typically associated with electron microprobes). Comparisons of the two systems have been made by numerous authors (refer to Birks, 1971; Reed, 1975; Geller, 1977; and Goldstein et al, 1981, for comprehensive comparisons); this chapter is devoted to the more popular energy-dispersive method.

## ORIGIN OF THE X-RAY SIGNAL

As described in Chapter 1, electrons inelastically scattered within the excitation volume of a specimen deposit some energy in many atoms. In order to return to ground state, the atom releases a distinct quantum of energy. If the excited atom ejects an inner-shell electron, an outer-shell electron fills that vacancy and emits an X-ray having an energy equal to the difference between the two electron shells. K-shell electrons are closest to the nucleus and therefore are more tightly bound than L-, M-, or N-shell electrons, which are progressively farther from the nucleus. For this reason, in a given atom K-shell electrons are more energetic than L-shell electrons, which in turn are more energetic than M-shell electrons. A given shell is then subdivided into $\alpha$ or $\beta$ levels, i.e., K$\alpha$, K$\beta$, L$\alpha$, L$\beta$, etc., and may be further subdivided into K$\alpha_1$, K$\alpha_2$, K$\beta_1$, etc., to accommodate the energy variations of the electrons within a shell.

The probability that a given transition will occur largely determines the intensity of the emission, and that probability increases as the distance between shells decreases. *K$\alpha$ X-rays* arise from L- to K-shell transitions and have the highest rate of emission, thus forming more prominent peaks than other transitions (Fig. 3-1). *K$\beta$ X-rays* arise from M- to K-shell transitions; because the distance between these shells is greater and the probability of this transition is lower than that of an L- to K-shell transition, K$\beta$ X-ray peaks are lower in height than K$\alpha$ X-ray peaks. K$\beta$ X-rays are of higher energy than K$\alpha$ X-rays, because the energy difference between K and M shells is greater than that between K and L shells. An obvious corollary to these statements is that a K$\beta$ X-ray will not be detected unless the more intense K$\alpha$ X-ray is also present.

The L lines arise from transitions between shells farther from the nucleus. *L$\alpha$ X-rays* arise from M- to L-shell transitions, and the *L$\beta$ lines* arise from N- to L-shell transitions. Both of these lines are of lower energy and intensity than K X-rays, and are usually detected for iron and heavier elements (the Fe L$\alpha$ line is at 0.704 keV). The L lines are used to identify elements heavier than zirconium, atomic number 40 (e.g., Mo, Ag, Cd, and Sn); for reasons to be discussed, the K lines of heavier elements are not excited during a routine analysis.

The final shell of interest to scanning electron microscopists is the *M shell*. The M$\alpha$ line arises from N- to M-shell transitions; these are the least energetic X-rays and are detected only from elements heavier than lanthanum (atomic number 57). In conjunction with the L lines, the common elements identified by their M-shell emissions are Ta, W, Pt, Au, Hg, Pb, and Bi. Figure 3-2 shows the relative energies of the KLM X-rays as a function of atomic number.

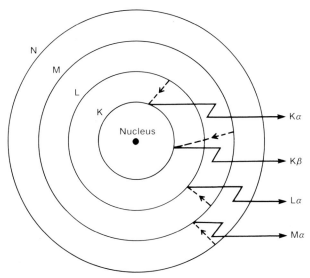

K shell
  Kα X-rays are the most intense X-rays and originate from L-shell →
    K-shell transitions
  Kβ X-rays are the most energetic X-rays and originate from M-shell →
    K-shell transitions
L shell
  Lα X-rays originate from M-shell → L-shell transitions
M shell
  Mα X-rays originate from N-shell → M-shell transitions

Energy levels: M < L < Kα < Kβ; intensity levels: Kα > Kβ > L > M

**Fig. 3-1.** The origin of X-rays as shown in the Bohr model of
the atom.

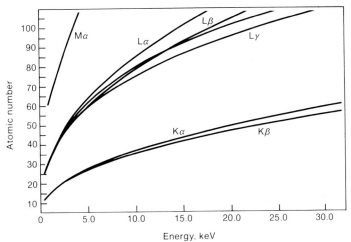

**Fig. 3-2.** The relation between atomic number and characteristic X-ray lines. (Courtesy of Dr. Nicholas Barbi and PGT)

For a given element, one may estimate the relative intensities of the different lines using the following chart where the reference point is $K\alpha_1 = 100$:

| K lines | L lines | M lines |
|---------|---------|---------|
| $K\alpha_1$ = 100 | $L\alpha_1$ = 100 | $M\alpha_{1,2}$ = 100 |
| $K\alpha_2$ = 50 | $L\alpha_2$ = 10 | $M\beta$ = 60 |
| $K\alpha_{1,2}$ = 150 | $L\beta_1$ = 50 | |
| $K\beta_1$ = 15-30 | $L\beta_2$ = 20 | |
| $K\beta_2$ = 1-10 | $L\gamma_1$ = 1-10 | |

A given shell will be excited only if the *binding energy* of its electrons is exceeded; i.e., the accelerating voltage must exceed the binding energy of the electron. *The critical excitation-edge energy* or *absorption-edge energy* refers to the minimal energy requirement for excitation of a given shell and is designated on X-ray tables as $K_{ab}$, $L_{ab}$, and $M_{ab}$. As a rule of thumb, the accelerating voltage should exceed the critical excitation energy by a factor of 1.5 to 3.0; because the highest accelerating voltage available on typical microscopes is 30 keV, the highest-energy X-rays detectable in EDS fall within the 10-20 keV range. The minimum energy detectable is roughly 0.7 keV and is a function of instrumental factors (discussed below).

To summarize, energy-dispersive spectrometers are capable of analyzing X-rays in the 0.7-20 keV range; except for elements with a very low atomic number ($<10$), all will release at least one X-ray within this range of energies. Even though it may not be possible to excite or detect all the characteristic X-rays of a given element, in general an element may be uniquely identified by the detectable lines.

The spectra of iron (atomic number 26) and lead (atomic number 82) are used as examples (Fig. 3-3). The characteristic X-rays of each element are listed below:

| Characteristic, lines | Element, keV Fe | Pb |
|------------------------|---------|---------|
| $M\alpha$.......... | – | 2.346 |
| $M\beta$.......... | – | 2.443 |
| $L\alpha_1$......... | 0.704 | 10.549 |
| $L\alpha_2$......... | – | 10.448 |
| $L\beta_1$.......... | 0.717 | 12.611 |
| $L\beta_2$......... | – | 12.620 |
| $L\gamma_1$......... | – | 14.762 |
| $K\alpha_1$ ........ | 6.403 | 74.957 |
| $K\alpha_2$ ........ | 7.057 | 72.794 |
| $K\beta_1$ ........ | – | 84.922 |
| $K\beta_2$ ........ | – | 87.343 |

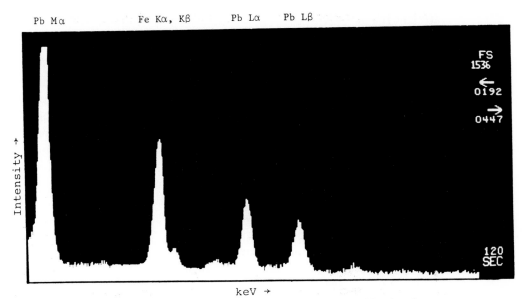

Fig. 3-3.  The characteristic spectra of lead and iron.

If each specimen is irradiated at 30 keV, all of the iron K and L lines will be excited, but only the M and L shells of lead will be excited: the K lines of lead will not be excited because the binding energy of these shells is beyond the energy limit of the SEM. Each element may be uniquely identified by the presence of peaks at the characteristic energies listed above. At this point, note that spectrometer resolution does not allow separation of very closely spaced peaks: for example, the iron $L\alpha_1$ and $L\beta_1$ form a single peak at 0.711 keV, which is the average of both peaks. This degradation of spectrometer resolution is discussed below.

In addition to characteristic X-rays, an irradiated specimen will emit *noncharacteristic X-rays* which are collectively referred to as *Bremsstrahlung, braking radiation, continuum,* or *white radiation*. Noncharacteristic X-rays are manifested as background in a spectrum because their energies are independent of elemental composition; they arise from deceleration of electrons which enter the nuclear field of an atom. The shape of the background follows an asymmetric curve that is highest at low energies (roughly 1-2.5 keV) and gradually slopes downward at higher energies. The intensity of the continuum therefore has a greater effect on lower-energy X-ray lines, and the same holds true when analyzing low- and high-Z specimens. Because low-energy peaks are sitting atop an already intense background level, they may appear more intense. In a quantitative analysis, the peaks are stripped of background and their true intensities analyzed. It is always desirable to suppress the background and enhance the characteristic X-ray signals (i.e., improve the signal-to-noise ratio); methods to suppress background are discussed later in this chapter.

## Absorption Effects

X-ray emission is affected by the atomic number (Z) of the specimen and by *self-absorption* of X-rays either by the emitting atom or by other atoms excited by interaction with the emitted X-rays to produce *X-ray fluorescence*. All of these factors are interrelated to the excitation volume (see Chapter 1): X-rays can originate from anywhere within the excitation volume with the bulk of the X-ray signal arising at a considerable distance from the specimen surface. The trajectory of primary-beam electrons and both the trajectory and origin of emitted X-rays are simulated using *Monte Carlo calculations* (Heinrich et al, 1976; Kyser, 1979). These are very rigorous mathematical treatments of X-ray data that quantify the X-ray volume, and are usually encountered only in research situations. Figure 3-4 gives Monte Carlo representations of the effect of atomic weight, accelerating voltage, and tilt on the trajectory of primary-beam electrons and secondary X-rays within specimens. These plots confirm that: (1) the excitation volume in low-Z materials is larger than in high-Z materials; (2) as accelerating voltage is increased, both the depth of penetration and the size of the excitation volume increase; and (3) the angle of incidence defines the relative position of the excitation volume much more than does the specimen surface (i.e., tilting the specimen does not alter the relative position of the excitation volume, but it does enhance counting statistics). The interrelationships of these parameters and their effect on an analysis are described below.

*Beer's Law* defines X-ray absorption as

$$\frac{I}{I_0} = \exp\left(-\frac{\mu}{\rho}\rho x\right)$$

where $I$ is intensity following absorption, $I_0$ is generated X-ray intensity, $\mu/\rho$ is mass absorption coefficient, $\rho$ is density of specimen (g/cm$^3$), and $x$ is thickness (cm). The *absorption coefficients* for pure elements have been calculated and may be found in X-ray energy tables, or they can be calculated from

$$\mu_M = kZ^3/E^3$$

where $\mu_M$ is the mass absorption coefficient, $Z$ is the atomic number, and $E$ is the energy of the incident X-ray. The *mass absorption coefficient* for a multi-element matrix is the summation of $\mu_M$ for all the contained elements:

$$(\mu_M) \text{ matrix} = \sum_i W_i(\mu_M)_i$$

where $W_i$ is the weight fraction of element $i$ in the matrix and $(\mu_M)_i$ is the mass absorption coefficient of element $i$ for the X-ray of interest.

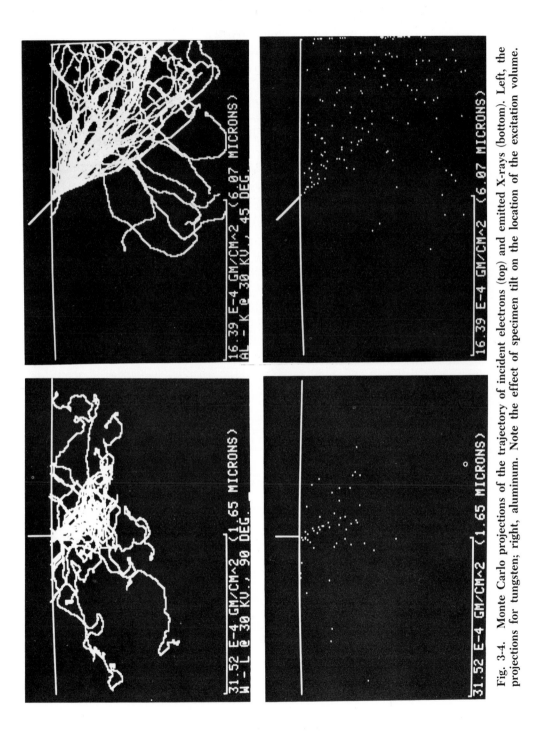

Fig. 3-4. Monte Carlo projections of the trajectory of incident electrons (top) and emitted X-rays (bottom). Left, the projections for tungsten; right, aluminum. Note the effect of specimen tilt on the location of the excitation volume.

The effects of absorption are thus manifested, for a given specimen matrix, on the basis of X-ray energy levels: as the energy level increases, absorption decreases. When voltage decreases, absorption increases until it peaks at the *K absorption edge* ($K_{ab}$). At the absorption-edge energy the X-rays are of sufficient energy to ionize other atoms. Below $K_{ab}$ the energy of the X-rays is below the binding energy of the electrons and absorption is again reduced. Clearly, absorption has a very real effect on EDS in terms of fluorescent yield, defined as the number of X-rays which escape from the sample, and X-ray fluorescence, a secondary emission arising from the excitation of low-energy X-rays by higher-energy X-rays.

The *fluorescent yield* ($\omega$) refers to those interactions that result in the emission of X-rays from a specimen. When X-rays are reabsorbed by an atom, the atom then releases an *Auger electron* of unique energy from an outer shell. The number of Auger electrons and X-rays emitted is inversely related to atomic number: $\omega$ increases with Z, and for a given element the *fluorescent yield* is K > L > M, while the yield of Auger electrons increases with decreasing atomic number (Fig. 3-5). The inverse relation between the quantity of emitted X-rays and Auger electrons implies that Auger spectroscopy is most useful for analyzing elements having very low atomic weight, whereas energy-dispersive spectroscopy is more suitable for elements heavier than (and including) sodium.

Because X-rays originate from a significant depth within the specimen, some X-rays will be absorbed by other atoms in the specimen. This self-absorption gives rise to *secondary emission* (or X-ray fluorescence), defined as the excitation of lower-energy shells by higher-energy X-rays.

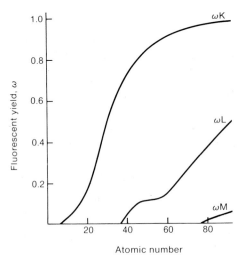

**Fig. 3-5.**   The fluorescent yield as a function of atomic number. (Courtesy of Dr. Nicholas Barbi and PGT)

This results in the artificial enhancement of low-energy peaks by the higher-energy peaks, which are suppressed as a result of the same mechanism. For example, if one were analyzing an iron-nickel alloy, the Ni K$\alpha$ lines (K$\alpha_1$, 7.477 keV; K$\alpha_2$, 7.460 keV) and Ni K$\beta$ lines (K$\beta_1$, 8.264 keV; K$\beta_2$, 8.328 keV) can enhance the intensity of the Fe lines. If a high-energy Ni X-ray strikes an Fe atom and the binding energy of an Fe K-shell electron is exceeded, the Fe atom releases that electron, and during its return to ground potential a K X-ray is released. This absorption-fluorescence phenomenon must be corrected when quantitative analyses are conducted. Several programs which accommodate these corrections have been developed; the *ZAF correction*, where Z = atomic number, A = absorption, and F = fluorescence, is a generic name for such operations.

In summary, two types of absorption influence the *fluorescent yield* (emission) of X-rays: (1) the self-absorption of an X-ray by the emitting atom, which then releases an Auger electron, is most prevalent with low-Z materials; (2) the absorption of an X-ray by an atom different from the emitting atom can cause secondary emission of lower-energy X-rays, which artificially enhances or suppresses peak intensity. These events are important when the specimen is composed of several elements but become less critical when pure elements are being analyzed. Clearly, both events must be corrected for in a quantitative analysis, but they may be discounted during a qualitative analysis.

# DETECTION AND PROCESSING OF THE X-RAY SIGNAL

## Instrumentation

The heart of an energy-dispersive spectrometer is a lithium-drifted silicon diode (semiconductor or solid-state detector) which detects emitted X-rays (Fig. 3-6). Abbreviated *Si(Li)* and pronounced "silly," the diode is composed of moderately pure silicon doped with lithium. This *drifted zone* contains an equal number of $Li^+$ donors and $B^-$ acceptors evenly distributed in the active area ($\sim 12.5$ mm$^2$) of the diode. The front of the diode is covered by a thin gold film capable of carrying a voltage. During operation, the diode is reverse biased by 1000 volts, which establishes a *depletion zone* by removing the normally present electron-hole pairs. An X-ray entering the depletion zone produces electron-hole pairs that are swept out by the bias and reappear as a pulse of charge on the opposite side of the diode. The magnitude of the pulse corresponds to the energy of the incident X-ray, and this proportionality is maintained during subsequent amplification, shaping, and display of the data. Further information on the properties of Si(Li) detectors is available in Woldseth (1973), Barbi and Lister (1981), and Fink (1981).

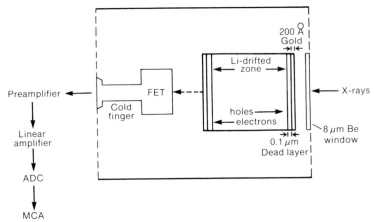

**Fig. 3-6.    Construction of the lithium-drifted silicon detector.**

Located directly behind the Si(Li) detector is the *field-effect transistor (FET)*, which is the first and most critical step in signal amplification. The FET is technically a preamplifier which serves to integrate the total charge of the pulse while converting it into a proportional voltage signal. Both the Si(Li) semiconductor and the FET are held at liquid nitrogen temperature via a *cold finger* to, respectively, prevent free diffusion of lithium through the semiconductor and reduce electronic noise. If a high level of noise were introduced at this stage, the noise would be amplified and a very poor signal displayed. The cold finger is cooled by connection to a *dewar flask* filled with liquid nitrogen; for obvious reasons the dewar should not be empty, either by intention or by accident, unless the high voltage to the diode has been removed.

The housing of the cold finger/detector assembly is referred to as the *end cap*. It is separated from the face of the diode by a very thin (~7.5 μm thick) *beryllium window*, which serves to protect the Si(Li) crystal from mechanical damage and ensure vacuum integrity. The window is of sufficiently low atomic number (4) to permit transmission of all but very low energy X-rays. *Windowless detectors* were introduced several years ago for light-element detection, but they are not usually found in service-oriented SEM facilities (e.g., refer to Barbi and Russ, 1975; Russ, 1977; Musket, 1981; Russ and Sandborg, 1981).

After leaving the preamplifier, the pulse, the height of which remains proportional to the energy of the incident X-ray, is further amplified and shaped in the main *amplifier*. The signal is electronically modulated in the amplifier in order to enhance the signal-to-noise ratio. Because amplification is a time-dependent function, "pulse pile-up" may occur if the count rate is excessive; the amplifier simply cannot process the pulses as rapidly as they are being generated. *Pulse pile-up*, a type of *dead time*, is defined as the probability that two or more X-rays will enter the detector nearly simultaneously, causing overlap of both signals. This event is manifested as a distortion of peak symmetry or the formation of a *sum peak*, the energy of which is the sum of the two peaks (the latter is discussed further under "Qualitative Energy-Dispersive Spectroscopy"). Most spectrometers are capable of rejecting pulse pile-ups, and the analyst can reduce the probability of this event by maintaining moderate count rates. The electronics employed during amplification are beyond the scope of this text; interested readers should consult Woldseth (1973) and Statham (1981) for comprehensive reports.

The shaped and amplified analog signal is converted into a digital signal by the *analog-to-digital converter* (ADC) and then sorted, stored, and displayed in a *multichannel analyzer* (MCA). Data may also be retrieved for qualitative or quantitative analysis. Data acquisition is a real-time event, and as each pulse enters it is sorted, by energy level, and entered into a channel which corresponds to that energy. The horizontal axis of the spectral display is scaled as channels or energy (keV), and the vertical axis represents counts. It is the number of counts at each energy level that is proportional to the composition of the specimen. As will be discussed, a significant number of counts are accumulated for statistical accuracy; the major peaks are identified by relating their energy to tables of characteristic energy, the minor peaks are identified (many of which are less intense lines of elements identified in the previous step), and the data are reduced by quantitative analysis, if desired. Modern multichannel analyzers conduct all of these operations; further information on the theory and operation of MCAs is thoroughly reviewed by Lifshin and Hayashi (1981).

## Spectrometer Resolution

*Spectrometer resolution* is a function of the number of channels that a peak encompasses (peak width) in a spectrum. X-ray spectra exhibit Gaussian-shape peaks rather than sharp lines. This is a result of signal degradation due to the statistical nature of electron-hole pair production in the Si(Li) detector and also due to the amplification and shaping of the X-ray signal. The resolution of a spectrometer is expressed as the *full width at half maximum (FWHM)*, usually for the 5.9-keV Mn K$\alpha$ peak from a radioactive Fe[55] source (Fig. 3-7). The FWHM is a function of noise ($N$) and statistical ($S$) contributions, and for the Mn K$\alpha$ peak may be simply expressed as

$$FWHM_{Mn K\alpha} = \sqrt{N^2 + S^2}$$

This value may be used to calculate the FWHM at any other energy using the following equation:

$$FWHM = \sqrt{R^2 + 2.623E - 15460}$$

where $R$ is detector resolution for the Mn K$\alpha$ peak in eV and $E$ is X-ray energy in eV.

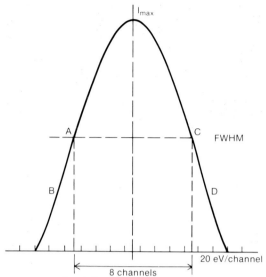

Fig. 3-7.   The full width at half maximum (FWHM).

Modern spectrometers can achieve a FWHM of $\cong 140$ eV; i.e., they can separate two peaks separated by a minimum of 140 eV. Two peaks having an energy difference less than 140 eV form one peak, the energy of which is the average of the two peak energies. For this reason, a typical spectrometer will combine the $L\alpha_1$ and $L\alpha_2$ lines to produce a single $L\alpha$ peak, and likewise, $K\alpha_1$ and $K\alpha_2$ may form one peak. For example, iron $K\alpha_1$ (6.403 keV) and $K\alpha_2$ (6.390 keV) form one $K\alpha_2$ peak at 6.40 keV.

*Peak overlap* has consequences for both quantitative and qualitative analyses. The characteristic X-rays of some elements overlap, and this may be a serious problem if the region of interest of a spectrum contains several closely spaced peaks. For example, consider the analysis of a cobalt-base superalloy, Haynes-25 (WF-11, L605; Fig. 3-8). This alloy is composed of the following metals, among others, and the most prominent X-ray peaks (rounded off) are indicated:

| Element, % | M | Significant peaks, keV | | | |
| --- | --- | --- | --- | --- | --- |
| | | $L\alpha$ | $L\beta$ | $K\alpha$ | $K\beta$ |
| Cr, 20.13%........ | – | – | – | 5.4 | 5.9 |
| Mn, 1.34 ......... | – | – | – | 5.9 | 6.5 |
| Fe,  2.18 ......... | – | – | – | 6.4 | 7.06 |
| Co, 51.31 ......... | – | – | – | 6.9 | 7.7 |
| Ni,  9.95 ......... | – | – | – | 7.5 | 8.3 |
| W, 14.75.........  | 1.8 | 8.4 | 9.8 | – | – |

In this situation, the $K\alpha$ of manganese will be hidden by the Cr $K\beta$ peak, and its $K\beta$ will be hidden by the iron $K\alpha$ peak. In a qualitative analysis one could crudely compare peak heights and determine, from

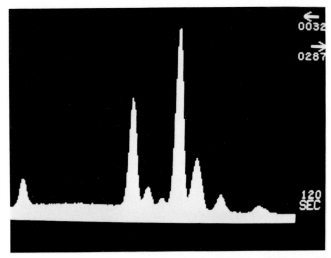

**Fig. 3-8.   Characteristic X-ray spectrum of Haynes-25.**

the inconsistent ratio of $K\alpha$ to $K\beta$, that manganese is present. However, if only a very low concentration of manganese were present, the analyst would not be able to identify it. In this spectrum also note that cobalt, nickel, and tungsten all have interfering peaks, but all may be positively identified by the presence of other characteristic X-rays and by the relative heights of the different peaks. If this specimen were an unknown, the analyst should be extremely careful, referring to it as belonging to the classification of cobalt-base alloys but not identifying it uniquely as Haynes-25 unless another analytical method was used for quantitative identification (e.g., atomic absorption spectroscopy).

Energy-dispersive spectrometers are most efficient for the analysis of X-rays having energies between ~0.7 keV and 15.0 keV. Spectrometer resolution at energies below ~0.7 keV is degraded by absorption of the X-rays in the beryllium window, by the gold thin film at the face of the diode, and by the thin inactive layer at the surface of the diode. Analysis of an element having a lower atomic number than sodium (11) is not considered possible. The Na $K\alpha$ peak is at 1.041 keV; roughly between 0.8 and 1.0 keV one will observe the L lines of moderate-weight elements (e.g., Fe, Cu, Zn). Windowless detectors eliminate absorption by beryllium and thus are capable of resolving very low energy peaks from carbon (atomic number 6; $K\alpha$, 0.282 keV) through aluminum (atomic number 13; $K\alpha$, 1.48 keV).

Beyond 15 keV, spectrometer resolution is diminished because these X-rays possess sufficient energy to completely pass through the Si(Li) detector. Also recall that many of these high-energy lines will not be excited because the SEM accelerating voltage is lower than the binding energy of those electron shells. The analytical energy range defined above should not be interpreted as a limitation of EDS. A large proportion of the periodic table will emit at least one analyzable X-ray that falls between 0.8 and 15 keV. This includes a very broad scope of specimen types, but usually limits the study of organic compounds.

# QUALITATIVE ENERGY-DISPERSIVE SPECTROSCOPY

A *qualitative analysis* is conducted to determine the elemental composition of a specimen, whereas a *quantitative analysis* reveals the relative concentrations of the elements detected. In either situation the data must be of high quality to ensure analytical confidence. The analysis is controlled by detector efficiency and spectral resolution, as discussed above, and also by SEM operating parameters, covered below.

## Spatial Resolution

The *spatial resolution* of an EDS analysis is analogous to the spatial resolution of images referred to in Chapter 1, and refers to the lateral dimensions of the volume representing the source of X-rays. Spatial resolution should not be confused with spectral resolution (full width at half maximum); the latter is a parameter uniquely defining the resolution of the spectrometer system. As noted earlier, the information depth encompasses a considerable volume beneath the specimen surface, and the bulk of X-rays originate from the deepest and broadest zone of the excitation volume (Fig. 1-5). The width of the X-ray excitation volume is the equivalent of spatial resolution: lateral-beam spread causes X-ray spatial resolution to be considerably poorer than the spatial resolution (~60 Å) of SEM images.

Although spatial resolution may be ignored when one is studying the bulk composition of a specimen, it is crucial when only a small fraction of the bulk sample is being analyzed, e.g., an inclusion. To conduct an analysis, one chooses a typical field of view, fills the observation CRT with that image (or places the beam "dot" at the desired position), and acquires a spectrum. Although a bulk analysis is usually conducted at low magnifications, once the image is magnified in order to resolve fine structures the excitation volume will change, but will still originate predominantly from beneath the specimen surface. To continue the inclusion example, one may magnify that image until it fills the field of view, but it is absolutely incorrect to assume that the acquired spectrum defines only the elemental composition of the inclusion. If the depth of penetration of the primary-beam electrons is greater than the depth of the inclusion, the beam will pass through the inclusion and excite the underlying matrix, which contributes to the X-ray signal. Consequently, one would prefer a backscatter image, which more closely approximates the X-ray volume than does a conventional electron image, and on this basis would choose a means to decrease the depth of penetration of the primary-beam electrons.* For this same reason, powdered or particulate samples are mounted on carbon planchets rather than aluminum stubs; the aluminum can be excited and emit its characteristic radiation, whereas carbon is beneath the resolution limit of conventional spectrometers.

X-ray spatial resolution may be enhanced by analyzing thin sections rather than bulk specimens, and by reducing accelerating voltage and spot size. In thin sections, lateral-beam spread is minimized simply because the dimensions of the specimens are small. Lowering the accelerating voltage is effective, provided that it is still above the critical excitation-edge energy. It may be desirable to conduct the analysis at two accelerating voltages to ensure that the minimum energy is exceeded. A reduced spot size is also effective, but only to the point where the depth of penetration is defined solely by the accelerating voltage. Clearly, lowering the accelerating voltage or reducing the spot size is only a remedial means of improving spatial resolution. Analysis of thin sections is the most effective technique, but not all specimens can be prepared in this manner. Consequently, the analyst should experiment with different operating conditions before drawing any conclusions about an unknown specimen.

Spot size and accelerating voltage also have a direct influence on the X-ray count rate. When the accelerating voltage is too low, characteristic X-rays will not be released because overvoltage has not been achieved; when the voltage is too high, the effects noted above predominate. A larger spot size increases the count rate and vice versa. Too large a spot size decreases resolution by increasing the noise level in the spectrum. Consequently, for a given specimen, the analyst should adjust spot size until a good count rate is achieved. Recall from Chapter 1 that the effective spot size may be increased by using large apertures; EDS analyses are normally conducted with apertures roughly 150-200 $\mu$m in diameter.

If the accelerating voltage and/or spot size are not optimum for a given specimen, two types of artifact peaks may be encountered (Fiori and Newbury, 1978 and 1980; Fiori et al, 1981). *Sum peaks* may be observed if the count rate is too high, and *escape peaks* are associated with too low an accelerating voltage. Sum peaks originate when two X-rays simultaneously enter the detector and their energies are added, forming another peak at that summed value. For example, the aluminum K$\alpha$ peak is at 1.487 keV; an aluminum sum peak would be roughly at $2(1.487) = 2.974$ keV or slightly below. The characteristic Al peak will be observed, and at excessive count rates the sum peak will also form.

---

*NOTE: If only a crude qualitative analysis were desired, one could analyze the bulk and compare that spectrum to the spectrum of the inclusion. If the inclusion were silicon-based, the Si peak heights would be significantly different, provided that other operating parameters were identical.

Sum peaks may be eliminated by reducing the spot size and/or the accelerating voltage. Escape peaks may be observed when elements heavier than sulfur (atomic number 16) are analyzed. The characteristic X-rays emitted by sulfur and heavier elements carry sufficient energy to excite the silicon $K\alpha$ line (1.74 keV) in the Si(Li) detector. The escape peak is manifested as a smaller peak 1.74 keV less than the major line. For example, an escape peak due to Fe $K\alpha$ excitation may be acquired at $6.403 - 1.74 = 4.663$ keV. The escape peak is always of lower intensity than the peak causing the excitation. Escape peaks are avoided by increasing the accelerating voltage.

To summarize, the analyst should feel comfortable with various operating conditions and be willing to experiment with different conditions for a given specimen. Evaluating the effects of different operating conditions does not require a significant amount of time, and the optimal set may be quickly chosen. Finally, the analyst should not choose the maximum accelerating voltage simply because it is available; whereas high voltages are desirable for analyzing high atomic weights (e.g., lead in paint samples), lower voltages are preferred for low-atomic-weight analyses (e.g., aluminum alloys).

## Geometrical Considerations

The beam-specimen-detector geometry also affects the outcome of an analysis. Because X-rays are of relatively low energy, it is always desirable to position the specimen in close proximity to the detector. Most X-ray detectors are mounted on a movable support, allowing the detector to be laterally positioned very close to the specimen. In addition, the specimen working distance should be adjusted with the Z-axis control until the specimen-detector distance is small. When examining metals or any specimen exhibiting a high proportion of backscattered electrons, note that these BSE may strike the pole piece of the final lens, giving rise to an artificial iron peak. In fact, BSE can rebound from any internal surface of the specimen chamber and cause emission of X-rays that do not originate from the specimen. Similarly, carbon planchets are used in place of aluminum stubs to minimize false aluminum peaks in aluminum-containing specimens.

The specimen angle (tilt) also affects the analysis, as discussed earlier. The important parameter to recall here is that the excitation volume induced during irradiation follows the pathway of the incident radiation and is essentially independent of the specimen surface. Given a low-Z specimen, the shape of the excitation volume will be an asymmetrical hemisphere relative to the sample surface (Fig. 3-9). This parameter becomes most influential when microvolumes are being analyzed; for example, when identifying inclusions within microvoids in a dimple-ruptured specimen, a variety of tilt angles should be employed to ensure that the X-ray signal is originating from the inclusion and not from the surrounding base metal.

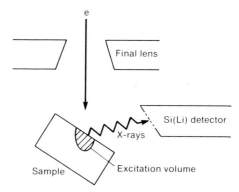

**Fig. 3-9.** The effect of specimen tilt on the location of the excitation volume.

Two additional geometrical parameters, the solid angle and the take-off angle, also affect an EDS analysis. The *solid angle* ($\Omega$) is the three-dimensional angle subtended at the sample by the detector. The *take-off angle* ($\psi$) is the angle between the sample surface and the line of sight to the center of the detector. Figure 3-10 identifies these angles relative to the sample and the detector.

While X-rays are generated in all directions during irradiation, only a small fraction are actually detected, and that fraction is defined by the solid angle. Woldseth (1973) offered the following analogy, which is an excellent description of the solid angle: imagine that the specimen is a sphere, the radius of which is the distance between the sample and the detector. The detector encompasses only a small fraction of that sphere.

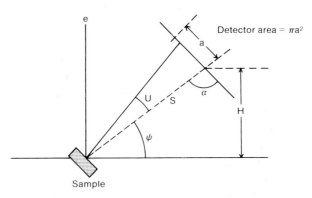

U = arc tan $\frac{a}{S}$

a = detector radius

S = distance from sample to detector

H = vertical distance from sample to center line of detector

Solid angle $\Omega = 2\pi (1 - \cos \mu) \sin \alpha \approx \frac{\pi a^2}{S^2} \sin \alpha$

Take-off angle $\psi$ = arc sin $\frac{H}{S}$

**Fig. 3-10.** Geometry of the take-off angle ($\psi$) and solid angle ($\Omega$). (Courtesy of Dr. Nicholas Barbi and PGT)

A conical-shape X-ray beam originating from the sample enters the detector: the angle of that cone is the solid angle. A high solid angle permits detection of a greater number of X-rays, and is achieved by using a small working distance. The take-off angle is defined as the angle between the sample surface and the line of sight to the center of the detector. A high take-off angle maximizes the number of X-rays that escape from the specimen, while a low take-off angle causes absorption of the X-rays by the sample. Therefore, the take-off angle should be $\geq 30°$ to ensure that the maximum number of X-rays will escape from the specimen and be detected.

## Data Display

X-ray data are displayed as spectra, line scans, or X-ray maps. Hard copies of the spectra are produced by recording a photograph of the MCA display screen or using a printer hard-wired to the MCA. Each peak is identified by positioning the MCA *cursor* or "bug" at the center of the peak, comparing that energy with tabulated characteristic X-ray energies, choosing the closest match, and reading the corresponding element. After the most intense peaks are identified, the less intense minor peaks can be read from the X-ray table. For example, if there is an intense peak at 6.40 keV, one refers to the X-ray tables to determine that this is the characteristic $K\alpha$ line of Fe; the table will indicate that iron also has a $K\beta$ peak at 7.05, and this less intense peak may be identified on the spectrum. Obviously, a $K\beta$ peak will not be detected unless the higher-intensity $K\alpha$ peak is also present. All remaining peaks are identified in the same manner.

If the energy levels of the observed peaks deviate by more than a few electron volts from those given in standard X-ray tables, the system is out of calibration. A simple *calibration standard* may be prepared using aluminum ($K\alpha$, 1.49 keV) and copper ($K\alpha$, 8.04 keV) tapes on a stub; because of the distance separating these peaks, the system can be calibrated according to manufacturer's specifications. Calibration is achieved when a newly acquired spectrum exhibits the center of the Al peak at 1.49 keV and the Cu peak at 8.04 keV. The calibration should be regularly checked and adjusted when necessary.

*Line scans* (synonym: line profile) and X-ray maps correlate the spectral display to the SEM image by feeding the X-ray data back into the microscope, resulting in the display of X-ray intensity. For example, if a cross section of a multilayered plating (each layer composed of a different element) is examined in the line-scan mode, the distribution of a given element will be displayed as a vertical deflection of the line scan. Each layer/element may be keyed into the SEM and all line scans simultaneously displayed. Such a display can reveal alloying between the plating layers.

*X-ray* or *dot maps* are also produced by feeding the X-ray data for a given element back into the SEM (McCarthy et al, 1981). The source of the X-rays is manifested as a grouping of extremely bright dots against a dark background where that element is absent. Dot maps are very useful for sorting inclusions of different composition and for mapping corrosion sites.

Regardless of how X-ray data are displayed, it is recommended that at least one electron image be recorded and that the site of the analysis be indicated on the micrograph. Ideally, both a secondary electron and backscattered electron image are recorded, and both images correlated with the X-ray data. This presents a much more complete package of information than a spectrum alone.

## QUANTITATIVE ENERGY-DISPERSIVE SPECTROSCOPY

Whereas a qualitative analysis reveals the elemental composition of a specimen, a quantitative analysis assigns relative concentration values to those elements. Quantitative analyses are possible because the intensity of the emitted radiation is proportional to the concentration of that element. Certified *reference standards* are available with very accurate compositional data from which one may "calibrate" a quantitative program and relate the observed intensity to the known concentration (i.e., using an empirical approach to the analysis). If it were not for the statistical nature of X-ray production, one could linearly relate intensity with concentration. Instead, the X-ray data are quantified for a given sample type, and the data are then compared with standards until the closest fit is achieved. More typically, however, the analyst is required to quantitatively analyze specimens for which a standard is nonexistent. So-called standardless programs are available in these situations, but they do not achieve the accuracy of programs using reference standards.

Sophisticated multichannel analyzers are capable of quantitative data reduction. Other multichannel analyzers may not be capable of such analyses, and data must be separately corrected. Older multichannel analyzers may be replaced with a hard-wired microcomputer that fulfills the same functions as the MCA and is sufficiently sophisticated to perform quantitative analysis.

The analyst must appreciate that the results of a quantitative analysis are only as good as the raw data, which presupposes that the SEM and EDS operating conditions selected by the analyst are optimal. These conditions, defined as those producing the best count rate, are a function of sample type, mean atomic weight, accelerating voltage, spot size, specimen tilt, take-off angle, working distance (both the Z-axis and lateral position of the X-ray detector), relative magnification, and the spectrometer characteristics described earlier in this chapter (also see Fiori and Newbury, 1978 and 1980; Barbi, 1980). When comparing a standard with an unknown specimen, the operating conditions should be identical to minimize any gross effect of the statistical nature of X-ray production (Ryder, 1975; Kotrba, 1977; Barbi, 1980). The author recommends that the analyst define a set of working conditions for various specimen types (e.g., polished samples, particles, etc); although these conditions may not be perfect for all specimens, they will serve as a general guide and will make it unnecessary to recalculate such factors as take-off angle for every new analysis. The raw and reduced data should be stored, along with the notation of operating parameters, and unknown spectra may be compared with the stored standard data until the best fit is achieved. Comparisons may be made under different operating conditions, but only rough qualitative comparisons may be made. Described below are methods to manipulate spectra and arrive at pure intensity data, which are then quantitatively analyzed through one of several matrix corrections. A number of different sample preparations are also briefly noted; these are discussed in greater detail in future chapters.

## Spectrum Manipulation

As a spectrum is acquired, characteristic peaks will accumulate counts as will the continuum, again because of the statistical nature of X-ray production. The noise level may be sufficiently high to conceal low-intensity and/or minor peaks. To enhance the clarity of the spectrum (i.e., improve the signal-to-noise ratio), the spectrum is *smoothed*. This does not improve quantitative data, but it improves the qualitative interpretation of a spectrum. The applications in which smoothing is most appropriate include the identification of closely spaced X-ray lines falling within the 2-4 keV range, and clarification of the peak centers in two overlapping lines. For example, the silver L$\alpha$ (2.98 keV) and L$\beta$ (~3.2 keV) normally cover a large number of channels with the two peaks distinct but overlapping; the analyst may smooth the spectrum and confirm that the centers of the L$\alpha$ and L$\beta$ peaks are characteristic in energy.

An alternative means of confirming the resolution of overlapping peaks is *spectrum stripping*. As described earlier, Haynes-25 superalloy exhibits overlap of several elements, the most severe overlap concealing the Mn K$\alpha$ by the Cr K$\beta$, and the Mn K$\beta$ by the Fe K$\alpha$. To resolve the manganese K$\alpha$ line, a pure chromium standard spectrum is stripped from the Haynes-25 spectrum to expose the Mn K$\alpha$ line. If desired, the characteristic iron peaks could also be removed by stripping a pure iron spectrum, revealing the Mn K$\beta$ peak. Stripping is very useful when several characteristic lines are hidden by more intense peaks. Stripping is more accurate than smoothing and is used for quantitative analysis.

Stripping is also very useful in distinguishing, for example, particles held within a filter from the filter itself. When small particles are analyzed, it is safe to assume that the filter will contribute to the spectrum. Because only the particle composition is of interest, one may strip a spectrum of the filter from the particle plus filter spectrum, thereby arriving at a spectrum of the particle alone. Spectral stripping is clearly a very useful method for both qualitative and quantitative analyses.

Because characteristic X-ray peaks form above a complex background, the determination of pure intensities requires that background be subtracted prior to data reduction. The continuum does not assume a linear shape in EDS spectra, and background subtraction is consequently a more sophisticated operation in EDS than it is in other systems. In EDS, background subtraction relies upon the definition of peak breadth and subsequent removal of frequencies other than the intermediate frequency assigned to true characteristic lines. Discarded are the low frequencies (continuum) and high frequencies (channel-to-channel statistical fluctuations), leaving only the pure intensities. Russ (1976) discusses in detail the various methods used for background subtraction.

Following background subtraction, the remaining peaks are integrated and the data presented in terms of intensity. The intensities are quantitatively analyzed and converted into concentration values through one of several matrix corrections.

## Matrix Corrections

The conversion of peak-intensity values into relative concentrations can follow several pathways, all of which are mathematically formidable. Because the analyst usually simply depresses a key to initiate these corrections, we will discuss only the general features of a quantitative analysis. Those readers desiring a more rigorous approach to this topic should consult the references cited for each method, as well as Yakowitz (1974), Goldstein et al (1977), Bolon (1978), Zaluzec (1979), Heinrich (1981), Myklebust et al (1981), and Russ (1981).

The presentation of relative concentrations normally involves calculation of the *k-factor*, which is the ratio between the unknown and standard X-ray intensities, along with corrections for absorption, fluorescence, and atomic number. Peak overlap and background subtraction are performed before the actual data reduction. The relation among these factors is defined by $C = kZAF$, where $k$ is intensity ratio, $Z$ is atomic number, $A$ is absorption, and $F$ is fluorescence. This seemingly simple relation is the basis for extremely complex data-reduction programs; an enormous number of factors, many of them in conflict, raise this equation to a very nontrivial level.

Quantitative programs which correct for atomic number $(Z)$, absorption by the sample matrix $(A)$, and fluorescence $(F)$ or secondary emission of X-rays are collectively known as *ZAF corrections*. Basically, these programs convert the raw spectral data into pure intensities for a given matrix either with or without reference standards. A practical example of a standardless application would be the classification of several paint samples into similar or different groups. Within this framework, a *ZAF* correction could reduce the raw data into concentrations and match similar samples. This type of operation is most useful if the qualitative spectra are similar.

The availability of standards enhances the data quality of a ZAF correction. For example, the asbestiform minerals may be subclassified into chrysolite, crocidolite, amosite, and anthophyllite. If standards of each subclassification were at hand (standards are available), one could specifically identify an unknown by comparison of its spectrum with the four standard spectra. Therefore, when available, standards are excellent starting points for any type of quantitative analysis and provide both increased accuracy and unique "fingerprints."

In addition to distinguishing quantitative programs on the basis of the presence or absence of standards, they may be classified on the basis of computer sophistication or sample type (e.g., mineral, particle, bulk, etc). *MAGIC* (Colby, 1968) is an extremely rigorous program that can be used only if the available computer memory exceeds 100 Kbytes, which is beyond the storage capacity of the computers usually interfaced with spectrometers. Colby (1971, 1978) subsequently introduced *MAGIC IV* and *MAGIC V*, which do not require such large-capacity storage. *FRAME C* (Myklebust et al, 1977, 1981) is another popular program used today for quantitative corrections. While the above programs may be applied to a variety of samples, especially bulk samples, other programs have been developed for unique sample configurations. Oxides and geological samples are analyzed using the *Bence-Albee Method* (Bence and Albee, 1968; Albee and Ray, 1970) while metal thin film data are quantified by the *Cliff and Lorimer Method* (Cliff and Lorimer, 1972; Goldstein et al, 1977).

Several of these programs are normally included in a quantitative software package. All of the manufacturers of these programs are excellent sources for the novice, who should be encouraged to evaluate several specimen types and the various corrections available. Continuing experience will simplify the analysis.

## Sample Preparation

**Metallographic mounts.** A polished sample is the ideal sample configuration and analytically the most accurate mode when a quantitative analysis is desired. Although metallography is usually associated with metals only, plastics, ceramics, and many other types of specimens may be prepared using exactly the same methods. With metals, note that the polished sample should not be etched for EDS, unless metallic inclusions or a unique phase, which may not be visible in the unetched conditions, is to be analyzed.

The sample should be finely polished, preferably to a mirror-like appearance. The final polish must be with diamond, rather than aluminum or silicon, to avoid surface contamination. The polished sample should be rinsed in distilled water followed by an organic solvent in an ultrasonic bath to remove any contamination. The mount is then placed in the SEM specimen stage, and because embedding media are poor conductors, a ground connection between the specimen and its holder is made with conductive paint or metallic tape. When using paint or tape, ensure that it does not contribute to the spectrum; simply use tape or paint that is not of the same composition as the specimen. If peaks arise from the pathway, move to another area of the sample.

When examining the edge of the specimen, it is difficult to avoid excitation of the mounting medium. It may be necessary to strip a spectrum of the medium from the characteristic spectrum of the specimen to eliminate interference. Another hint is to visibly position the specimen such that the surface of interest is parallel to the horizontal axis of the SEM observation screen; charging artifacts may be observed above or below the specimen, but not going through it. If the specimen itself is also nonconductive, charging may seriously distort both the SEM and EDS analysis. Charging may be minimized by evaporating a thin film of carbon across the surface. Carbon is a moderately good conductor and also of low atomic weight; characteristic X-rays will pass through the carbon film without a loss of energy. Do not coat nonconductive specimens with metal films when an EDS analysis is desired; metal films absorb low-energy X-rays, decrease the intensity of characteristic peaks, and emit their own characteristic lines, all of which are undesirable events. The detailed preparation of metallographic mounts, polishing, and thin films are discussed in later chapters.

**Powders.** A homogenous, loose powder may be qualitatively analyzed by mounting it on a carbon substrate. Simply draw a stripe of carbon paint across the carbon planchet, wait until the paint becomes tacky, and sprinkle a small amount of the powder across the paint. Excess particles are dislodged by tapping the planchet. The specimen is dried and analyzed. A better method for quantitative analysis is to compress the powder into a small aluminum cap. The pressed powder must be homogenous, uniform, and flat-surfaced. Pressed powders rather than discrete particles are preferred for quantitative analysis because the exact X-ray take-off angle may be calculated.

**Particles.** Airborne or liquid-borne particles are effectively concentrated by filtration. A section of the filter is removed and mounted on a carbon planchet, and discrete particles are analyzed. If the sampling is composed of several types of particles, many should be surveyed until the analyst can accurately assess the population. In this situation, it is extremely important that low filter loadings be used; the overlap of particles can lead to undesirable secondary fluorescence. Only individual particles should be analyzed. A background spectrum of the filter should also be recorded and subtracted from the spectra of the particles (see Small et al, 1978, and Statham and Pawley, 1978, for a novel approach to quantitative particle analysis). Finally, because filters are usually nonconductive, a carbon coating may be desirable. The thin film will also envelop the particles and diminish specimen loss.

**Rough specimens.** In the event that a rough surface (e.g., a fracture surface) is to be quantitatively analyzed, the microscopist should report results with a lower than normal level of confidence because the take-off angle for a given field of view is unknown. It may be worthwhile to alter the specimen tilt angle and accelerating voltage, and evaluate two sets of data for more valid results. The specimen surface must be free of any contaminants, which include hydrocarbons and dust. Specimen cleaning is discussed in Chapters 4 and 6.

A more detailed description of unique methods of sample preparation for quantitative analysis is available in Jenkins and DeVries (1970). Note that these methods are covered in several subsequent chapters.

## REFERENCES

Albee, A. L., and L. A. Ray (1970) Correction factors for electron probe microanalysis of silicates, oxides, carbonates, phosphates, and sulphates. *Anal. Chem.* 42:1408.

Barbi, N. C. (1980) Detectability in energy dispersive microanalysis. *SEM, Inc.* 2:297.

*_____ (1981) *Electron Probe Microanalysis Using Energy Dispersive X-ray Spectroscopy.* PGT, Inc., Princeton, NJ.

Barbi, N. C., and D. B. Lister (1981) A comparison of silicon and germanium X-ray detectors. In: *Energy Dispersive X-ray Spectrometry*, NBS Special Pub. 604. U.S. Govt. Printing Office, Washington, DC, p 35.

Barbi, N. C., and J. C. Russ (1975) Analysis of oxygen in an SEM using a windowless energy dispersive X-ray spectrometer. *IITRI/SEM*, p 85.

Bence, A. E., and A. L. Albee (1968) Empirical correction factors for electron microanalysis of silicates and oxides. *J. Geol.* 76:382.

Birks, L. S. (1971) *Electron Probe Microanalysis*, 2d ed. Wiley-Interscience, New York.

*Bolon, R. B. (1978) How to use quantitative X-ray analysis programs. *SEM, Inc.* 1:813.

Cliff, G., and G. W. Lorimer (1972) Quantitative analysis of thin metal foils using EMMA-4—the ratio technique. *Inst. of Physics (London):EM* 1972, p 140.

*Colby, J. W. (1968) MAGIC—a computer program for quantitative electron microprobe analysis. *Adv. X-ray Anal.* 11.

_____ (1971) MAGIC IV—a new improved version of MAGIC. *Proc. 6th Natl. Conf. Electron Probe Soc.*, 17A-B.

_____ (1978) MAGIC V—a computer program for quantitative electron excited energy dispersive analysis. Kevex Corp. Foster City, CA.

Fink, R. W. (181) Properties of silicon and germanium semiconductor detectors for X-ray spectrometry. In: *Energy Dispersive X-ray Spectrometry*, NBS Special Pub. 604. U.S. Govt. Printing Office, Washington, DC, p 5.

*Fiori, C. E., and D. E. Newbury (1978) Artifacts observed in energy dispersive X-ray spectrometry in the scanning electron microscope. *SEM, Inc.* 1:401.

_____ (1980) Artifacts in energy dispersive X-ray spectrometry in the scanning electron microscope (II). *SEM, Inc.* 2:251.

Fiori, C. E., et al (1981) Artifacts observed in energy dispersive X-ray spectrometry in electron beam instruments — a cautionary guide. In: *Energy Dispersive X-ray Spectrometry*, NBS Special Pub. 604. U.S. Govt. Printing Office, Washington, DC, p 315.

Geller, J. D. (1977) A comparison of minimum detection limits using energy and wavelength dispersive spectroscopy. *IITRI/SEM* 1:281.

Goldstein, J. I., et al (1977) Quantitative X-ray analysis in the electron microscope. *IITRI/SEM* 1:315.

_____ (1981) *Scanning Electron Microscopy and X-ray Microanalysis*. Plenum Press, New York.

Heinrich, K. F. J. (1981) *Electron Beam X-ray Microanalysis*. Van Nostrand Reinhold, New York.

Heinrich, K. F. J., et al, eds. (1976) *Monte Carlo Calculations in Scanning Electron Microscopy and Electron Probe Microanalysis*. NBS Spec. Pub. 460. U.S. Govt. Printing Office, Washington, DC.

Jenkins, R., and J. L. DeVries (1970) *Practical X-ray Spectrometry*. Springer-Verlag, New York, ch 8.

Kotrba, Z. (1977) The limit of detectability in X-ray electron-probe microanalysis. *Microchim. Acta* 2:97.

Kyser, D. F. (1979) Monte Carlo simulation in analytical electron microscopy. In: *Introduction to Analytical Electron Microscopy.* (Hren, J. J., et al, eds.) Plenum Press, New York, p 199.

Lifshin, E., and S. R. Hayashi (1981) Understanding multichannel analyzers. In: *Energy Dispersive X-ray Spectrometry*, NBS Special Pub. 604. U.S. Govt. Printing Office, Washington, DC, p 165.

McCarthy, J. J., et al (1981) Acquisition, storage, and display of video and X-ray images. *MAS Proc.*, p 30.

Musket, R. G. (1981) Properties and applications of windowless Si(Li) detectors. In: *Energy Dispersive X-ray Spectrometry*, NBS Special Pub. 604. U.S. Govt. Printing Office, Washington, DC, p 97.

Myklebust, R. L., et al (1977) FRAME C: A compact procedure for quantitative energy-dispersive electron probe X-ray analysis. *Proc. 12th Conf. Microbeam Analysis Soc.*, 96A-D.

_____ (1981) Spectral processing techniques in a quantitative energy dispersive X-ray microanalysis procedure (FRAME C). In: *Energy Dispersive X-ray Spectrometry*, NBS Special Pub. 604. U.S. Govt. Printing Office, Washington, DC, p 365.

*Newbury, D. E. (1979) Keynote paper: Microanalysis in the scanning electron microscope: Progress and prospects. *SEM, Inc.* 2:1.

Reed, S. J. B. (1975) *Electron Microscope Analysis*. University Press, Cambridge, Eng.

Russ, J. C. (1976) Processing of energy dispersive X-ray spectra. *EDAX EDITor* 6(3):4.

_____ (1977) Procedures for quantitative ultralight element energy dispersive X-ray analysis. *IITRI/SEM* 1:289.

_____ (1981) Multiple least squares fitting for spectrum deconvolution. In: *Energy Dispersive X-ray Spectrometry*, NBS Special Pub. 604. U.S. Govt. Printing Office, p 297.

*Russ, J. C., and A. O. Sandborg (1981) Use of windowless detectors for energy dispersive light element X-ray analysis. In: *Energy Dispersive X-ray Spectrometry*, NBS Special Pub. 604. U.S. Govt. Printing Office, p 71.

Ryder, P. L. (1975) Sensitivity, detectability limit, and information content of electron probe microanalysis. *IITRI/SEM*, p 111.

Small, J. A., et al (1978) The production and characterization of glass fibers and spheres for microanalysis. *SEM, Inc.* 1:445.

Statham, P. J. (1981) Electronic techniques for pulse-processing with solid-state X-ray detectors. In: *Energy Dispersive X-ray Spectrometry*, NBS Special Pub. 604. U.S. Govt. Printing Office, Washington, DC, p 141.

Statham, P. J., and J. B. Pawley (1978) Method for particle X-ray microanalysis based on peak-to-background measurements. *SEM, Inc.* 1:469.

Woldseth, R. (1973) *X-ray Energy Spectrometry*. Kevex Corp, Burlingame, CA.

Yakowitz, H. (1974) X-ray microanalysis in scanning electron microscopy. *IITRI/SEM*, p 1029.

Zaluzec, N. J. (1979) Quantitative X-ray microanalysis: Instrumental considerations and applications to materials science. In: *Introduction to Analytical Electron Microscopy*. (Hren, J. J. et al, eds.) Plenum Press, New York, p 121.

---

*Recommended reading.

# PART 2:
## Specimen Preparation

# =4=

# Introduction to Sample Preparation

The microscopist is challenged to characterize a variety of sample types and should be familiar with various methods of sample preparation. Outlined in this chapter are several methods which apply to a broad variety of samples. Each method is more thoroughly discussed in subsequent chapters.

Before beginning any preparation method, the analyst should examine the specimen with the naked eye and, if required, a low-power binocular microscope. General features should be noted and a sketch or photograph of the specimen recorded. When large specimens are to be studied, a sketch or photo is invaluable for orientation of the sample in the SEM chamber. Because the minimum magnification of most SEMs is roughly 10×, only a small portion of the specimen surface may be visible, and the microscopist may waste valuable time searching for the desired feature or features. Indicating the location of these features on a sketch or photograph provides a complete macroscopic and microscopic presentation that is much more valuable than SEM micrographs alone.

The analyst should be aware of the objectives of a given study before beginning the sample preparation. Especially when the specimen is nonconductive, different preparation methods are used depending upon whether an X-ray or straight SEM analysis is desired. In some situations using an inappropriate preparation may be disastrous, and the need to outline objectives becomes clear.

## SAMPLE SIZE

A very obvious criterion of sample preparation is that the sample cannot exceed the size of the specimen chamber. A corollary is that if the specimen fits into the chamber, the X, Y, Z, tilt, and rotate translations of the stage should not be severely limited. SEM manufacturers have responded to the needs of users by designing large-capacity specimen chambers which can accommodate large (~3 in.$^3$) specimens. These large chambers are useful when it is either impossible or undesirable to reduce the sample size. For example, in the microelectronics industry large wafers are often evaluated in the SEM; a size reduction would undermine the objective of the test. In failure analyses involving litigation, reducing the size of an exhibit may be viewed as tampering with the evidence. In both such situations a large specimen chamber may be essential.

Situations will nevertheless arise in which the specimen size exceeds the capacity of any chamber now available. A nondestructive method which circumvents this problem is the exact reproduction of the surface of interest using replicating material. Materials used for replicas include cellulose acetate tape and dental impression materials. When applied to the original surface either in a solvent (cellulose acetate softened with acetone) or as an elastic monomer (dental media) and allowed to cure, the replica will strip off as an exact but negative duplicate of the contact surface. The replica is then coated with a metal thin film and examined in the SEM. This extremely useful technique makes it possible to study any specimen which is too large to fit in the SEM chamber.

## SAMPLE CLEANING

Any debris on a specimen surface interferes with good imaging and X-ray analysis. In the SEM, contamination is manifested as areas of charging caused by the presence of nonconductive particles or films which obscure the sample surface. In addition to causing imaging problems and spurious X-ray signals, contaminants (especially hydrocarbons) sublime during vacuum exposure and deposit on the detector, apertures, column liner tube, etc. These problems are avoided by cleaning the specimen with organic solvents (e.g., acetone, ethanol, methanol, or a mixture) in an ultrasonic bath, followed by a blast of compressed gas. If the specimen cannot be solvent-cleaned (some plastics are soluble in organic solvents), it may be gently brushed with a soft camel's hair brush followed by a blast of compressed air. After cleaning, the sample should be carefully stored to avoid contamination by settled dust.

Oxidized metal samples require more rigorous cleaning. Any cleaning technique should be approached with caution and common sense, because the method chosen must not affect the base metal. For example, fracture surfaces often exhibit very fine structures such as fatigue striations or microvoids which could easily be destroyed by an aggressive cleaning method. The least aggressive cleaning methods include air dusting and solvent cleaning, while the most aggressive treatment employs acids containing corrosion inhibitors. A moderately aggressive but very effective cleaning method involves stripping several cellulose acetate replicas from the fracture surface. These "extraction replicas" will remove heavy oxide deposits and any other debris while the native fracture surface is inert. As a forewarning, it should be understood that oxidation consumes part of the fracture surface, and thus replica stripping does not restore the original surface, but provides more information than does the oxidized surface. The cleaning methods appropriate for different sample types are discussed in Chapter 6.

## SAMPLE MOUNTING

After cleaning, the specimen is mounted on a substrate which can be secured in the SEM specimen stage. Aluminum or carbon stubs are the standard specimen supports and are available in several sizes and configurations, depending upon SEM manufacturer. The manufacturer may also provide several types of specimen holders, e.g., one accepting stubs and another accepting standard-size metallographic mounts. Many microscopists improvise or design holders to fit their unique needs. For example, simple aluminum disks ~5 cm in diameter and ~3 mm thick are easily machined and are useful substrates for bulky samples.

Sample stability is extremely important during an SEM analysis; any movements other than the normal motions used to manipulate the sample are undesirable because a given field of view will shift, interfering with photography and EDS. The adhesives used to secure the specimen to a substrate are metallic tapes (aluminum or copper), double-sided tape, and conductive paints (silver or carbon); Murphy (1982) recently reviewed many adhesives, including tapes and paints. Unless the specimen is to be coated with a conductive thin film, double-sided tape or glue should be avoided; both are insulators and are prone to heating, which can result in specimen drift. Metallic tapes are useful for securing bulky or awkwardly shaped specimens, provided that the tape does not obscure the surface of interest or interfere with EDS. On large samples the metallic tapes may be used by the microscopist to define the regions of interest; by carefully positioning the tape and using "marker" arrows cut from the tape, the desired field of view and specimen orientation can be quickly found.

Metallic or double-sticky tapes have two major disadvantages: the adhesives will outgas during vacuum exposure, contaminating the SEM, and the adhesives per se are nonconductive. To avoid these problems, conductive paints are also used as adhesives. Silver or carbon paints are available, the carbon paint used when an EDS analysis is required (silver paint could release characteristic X-rays), and silver paint for any other type of analysis. Both paints are excellent adhesives that provide electrical contact to ground and minimal volatilization during evacuation. To avoid contamination of the sample by the capillary action of the liquid, let the paint set slightly (~10 sec) before positioning the sample. The paint must be dry before the specimen is exposed to vacuum to avoid microscope contamination.

## CONDUCTIVE COATINGS

Conductive thin films (100-250 Å thick) increase the density and conductivity of nonmetallic specimens which would otherwise act as beam absorbers during irradiation. Uncoated materials such as plastics, ceramics, glasses, and replicas are poor conductors that exhibit charging artifacts under normal operating conditions. Such specimens may be examined at low accelerating voltages, but magnification and resolution are limited. The application of a thin film to these surfaces significantly increases the maximum magnification and resolution limits.

Gold is the metal most commonly used for coating, and conductive thin films are prepared using either the evaporative (vacuum bell jar) or sputter coating technique. Gold or other metal films are used when imaging is desired. Carbon thin films are employed when EDS is desired; carbon films are prepared in vacuum bell jars. The carbon film is transparent to X-rays emitted by the sample, and does not contribute to the spectrum because carbon is of very low atomic weight. Metal films would both absorb low-intensity or low-energy X-rays and release characteristic peaks. Sputter and evaporative coatings are discussed in Chapter 8.

## PARTICLE ANALYSIS

The microscopist should be familiar with the techniques used to handle small particles, which include environmental particles, powders, abrasives, pigments, and wear debris. Because particle analysis is itself a "technology," discussed here are the very general methods used for studying particles. Interested readers should consult Beddow (1980) or other textbooks on particle technology for more information.

Since the surfaces of aluminum stubs are normally rough, particles tend to fall between the machining marks, making analysis difficult. Therefore, particles should be mounted for SEM analysis on very smooth substrates such as a glass microscope slide or coverslip, a polished stub, or a polished carbon planchet. Slides and coverslips are convenient only if the particles are to be coated with a conductive film. Polished aluminum stubs are useful for imaging but cannot be used when EDS is required because aluminum may contribute to the spectrum. Carbon planchets or stubs are ideal for combined EDS and SEM, because the particles alone release a detectable X-ray signal. The particles are glued to the stub using carbon paint. This method is most suitable for homogeneous, dry powders.

Particles suspended in a liquid are filtered to both concentrate and isolate the solid phase. The filter loading should be low enough to ensure that particle overlap is minimal; good SEM and EDS require the analysis of discrete particles. Not all filter types are suitable for SEM analysis; one prefers a filter that retains all particles at the surface rather than within the filter matrix. For example, fibrous filters (e.g., common paper, glass fiber, or cellulose ester membrane filters) entrap particles within the fibrous matrix, and only a small fraction of the total collection remains on the filter surface. Nuclear pore filters are preferred for SEM because all of the particles collected remain on the surface, which is itself very smooth. The pores of these polycarbonate filters are straight, individual, and cylindrical, and of variable size (0.2-0.8 $\mu$m for SEM).

The intercepted particles may then be examined by removing a section of the filter and coating with a metal or carbon film for, respectively, imaging and EDS. For EDS, a carbon substrate is also recommended. The particles may then be analyzed.

## REFERENCES

Beddow, J. K. (1980) *Particulate Science and Technology*. Chemical Publishing Co., New York.

Murphy, J. A. (1982) Consideration, materials, and procedure for specimen mounting prior to scanning electron microscope examination. *SEM, Inc.* 2:657.

# 5

# Polished Samples

The practice of metallography has evolved from an art into a science. This is largely due to the introduction of automatic polishing techniques, which provide reproducible polishing conditions with increased productivity, and to the standardization of etchant solutions. Polishing and etching reveal the microstructure of a given specimen as studied with the reflected light microscope, and further microstructural and compositional information may be gained using the SEM. Because metallography is a technology itself, the author has briefly outlined the general procedures but has intentionally omitted specific methodologies. Several excellent publications cited throughout this chapter give full justice to metallographic theory and practice; the novice metallographer should be familiar with at least one of these sources.

The SEM is not a replacement for the reflected light microscope: the two microscopes reveal different types of information and thus provide complementary data. A polished specimen must always be observed and characterized with the light microscope before an SEM analysis is conducted. Light photomicrographs may then be used as a guide during the SEM examination, much as macrophotographs are used with other types of specimens. When used in conjunction with each other, both types of optical instruments are excellent analytical tools (see Gnizak, 1982, for a discussion of intermicroscope correlations).

A variety of specimen types may be prepared as polished samples. In addition to metals (Davidson, 1981; Hall, 1981), appropriate specimens include plastics (Linke and Kopp, 1981), silicon crystals (Slepian et al, 1981), silicon chips (Liebl, 1981), coal (Stanton and Finkelman, 1979), and ceramics (Arrowsmith et al, 1978; Kestel, 1981; Weidmann, 1982). SEM is useful for examining polished samples in a variety of applications, including:

1. Single or multilayered platings for the evaluation of plating composition, continuity, dimensions, and the presence or absence of alloying between layers.

2. SEM imaging of extremely fine-grained microstructures which are too small to be resolved by the light microscope (e.g., very fine ferrite and pearlite in high-strength, low-alloy steels).
3. Identification of corrosion products in samples exhibiting stress-corrosion cracking.
4. Elemental identification of the base matrix, segregated phases, and inclusions in metals and ceramics.
5. Evaluation of the fine structure of hard filler in plastics.
6. Measurements of small air pores in concrete sections.

Outlined below is the general sequence of sample preparation (Fig. 5-1).

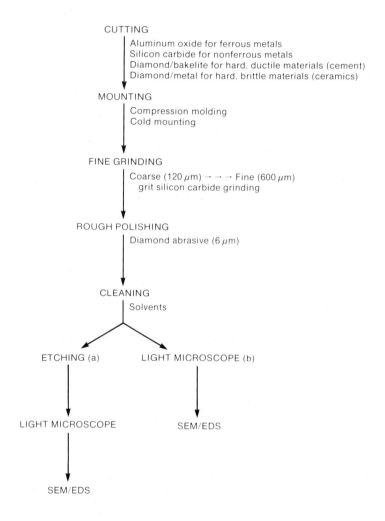

CUTTING
  Aluminum oxide for ferrous metals
  Silicon carbide for nonferrous metals
  Diamond/bakelite for hard, ductile materials (cement)
  Diamond/metal for hard, brittle materials (ceramics)

MOUNTING
  Compression molding
  Cold mounting

FINE GRINDING
  Coarse (120 $\mu$m) → ⋯ → Fine (600 $\mu$m)
    grit silicon carbide grinding

ROUGH POLISHING
  Diamond abrasive (6 $\mu$m)

CLEANING
  Solvents

ETCHING (a)          LIGHT MICROSCOPE (b)

LIGHT MICROSCOPE          SEM/EDS

SEM/EDS

(a) For microstructural analysis.
(b) Analysis of nonmetallic inclusions, cracks, pits, and intergranular corrosion.

**Fig. 5-1.   Flow chart for the preparation of metallographic mounts.**

# SAMPLE PREPARATION

Specimens that are too large to conveniently handle are cut into smaller sections. A sketch should be prepared showing the orientation of the section relative to the specimen, especially when the specimen has a complex geometry. The section to be removed is outlined with waterproof ink, and the method of cutting is chosen based upon the characteristics of the specimen (Nelson and Westrich, 1974; Leco, 1977; Buehler, 1981a; Struers, 1982). The cutting method must not affect the microstructure of the specimen or deform the cut surface. A variety of cut-off machines are commercially available, as are cut-off wheels of different abrasives (Fig. 5-1). For example, aluminum oxide wheels are used for cutting ferrous alloys and silicon carbide for cutting nonferrous alloys.

The sections are then compression-molded or cold-mounted to contain the sample in a standard cylindrical mold. The mold also protects the specimen edge from deformation during polishing. Specimens that are unaffected by heat and pressure are *compression-molded;* specimens that are sensitive to either heat or pressure are *cold-mounted* (Nelson, 1976; Nelson and Albrecht, 1976a, 1976b). Edge retention is particularly important when examining platings or any other surface-related structures. Edge distortion arises from shrinkage of the molding material during curing or from polishing an extremely hard specimen held within a softer mounting medium. Different media formulations are available that minimize or eliminate edge distortion, and these should be used when required.

The mount is then ground and polished to a mirror-like finish (Leco, 1977; Buehler, 1981b). Stepwise fine grinding from large grit (50 grit), through several intermediate moderate grits, to fine grit (600 grit) silicon carbide removes any deformations introduced during specimen cutting and produces a very flat surface. *Grinding* is usually accomplished with grinding wheels, with a moderate amount of pressure applied to the mounted specimen while the rotating wheel supplies the necessary motion. Each stage of fine grinding eliminates the distortions introduced by the previous grinding step, until at the final stage the specimen surface is flat and relatively featureless.

Rough and fine polishing produce a mirror-like surface that is free from distortions or unidirectional scratches. *Diamond abrasives* (~600 $\mu$m) are used for rough polishing, and the final polish usually employs *aluminum abrasives* (0.05-0.3 $\mu$m). Since very little material is actually removed during polishing, the preceding grinding steps must be thorough. Polishing may be performed using manual polishing wheels, although automatic polishers offer much more reproducible results and flatter surfaces (Dillinger, 1982; Samuels, 1982; Struers, 1982).

The polished mount is then thoroughly cleaned to remove any adhering grit or debris. Usually a rinse in warm water followed by methanol and quick air drying with a warm blow dryer is sufficient to prevent staining or corrosion of the sample. Turner and Rivenburgh (1979) recommend a mixture of 48% Freon, 48% methylene chloride, and 4% ethanol applied as a pressurized spray for cleaning mounts. This mixture and method of application thoroughly cleans both the polished surface and the interface between the sample and the mounting medium. An ultrasonic treatment using an organic solvent (or mixture) may also be used. Be careful if the specimen exhibits corrosion products; too vigorous a cleaning may displace the products.

A polished, unetched sample may reveal cracks, pits, intergranular corrosion, and nonmetallic inclusions that can be analyzed by light microscopy followed by SEM. Little or no information, however, is revealed about the specimen's microstructure unless it has been etched. *Etching* lends contrast to the microscopic image and usually involves preferential attack of the polished surface such that a very fine bas-relief of the microstructure results. The choice of an etchant is based largely upon the alloy composition and the particular feature sought; e.g., different etchants may be used to reveal different phases of a given alloy. Petzow (1978) comprehensively discusses etchants; the novice may prefer to consult a more general source (Leco, 1977), which covers a wide spectrum of common etchants and metals. The microstructure is then examined in the reflected light microscope. It may be difficult or impossible to resolve very fine martensite found in high-strength steels using the light microscope, but the resolution of fine microstructural features is straightforward in the SEM. Thus, the SEM is a useful secondary tool for microstructural analysis as well as providing elemental analysis capabilities (e.g., Johari et al, 1969; Kotval, 1969).

In addition to the references cited above, several excellent textbooks are available which comprehensively discuss metallographic sample preparation. The novice should consult at least one of the following references prior to beginning a metallographic study: American Society for Metals (1973), Goodhew (1973), Richardson (1974), McCall and Mueller (1974), McCall and French (1978), and Buhler and Hougardly (1980). The interpretation of microstructure can be found in American Society for Metals (1972), McCall and French (1977), and Samuels (1980). The preparation of plastics is described by Linke and Kopp (1981), that of ceramics by Kestel (1981) and Weidmann (1982), and silicon by Slepian et al (1981). The suppliers of metallographic equipment (Appendix C) are also excellent sources who are very willing to share their expertise.

## SEM OF POLISHED SAMPLES

As with any type of sample, polished mounts must be clean before beginning an analysis. The solvent cleaning methods discussed above are very effective, and settled dust is easily removed with compressed air. The specimen is then examined in the light microscope, and areas requiring SEM analysis are identified. Several photomicrographs, from low magnification to the desired magnification level, are recorded. The general location of the desired site may be indicated by placing one or two metal arrows (cut from metal tape) directly on the mount. The specimen is then securely mounted in an SEM specimen holder and a connection to ground established by painting a narrow stripe of silver paint (or tape) across the mount surface to the holder. Specimen holders uniquely for metallographic mounts are commercially available or may be machined from brass or aluminum. If a nonconductive specimen (plastic or glass) has been coated with a thin film, a dab of conductive paint between the mount perimeter and specimen holder is necessary as a ground connection.

Conductive thin films are usually applied only to those specimens that are completely nonconductive, such as ceramics or plastics. Carbon films are used for EDS, and metal films for microstructural evaluation. Metallic specimens may be coated with a carbon film to minimize charging of the mounting medium, but usually the ground connections mentioned above are adequate. When the edge of a metallic specimen is under study, a thin film may be desirable simply because the source of charging is very close to the area of interest.

The specimen is mounted in the SEM and the desired field of view is located. Focusing may be difficult because the specimen is smooth and flat; because the plane of the metal arrow is only slightly different from the specimen plane, coarse focusing the edge of the arrow at low magnification and then fine focusing the specimen at higher magnification is helpful. A large spot size is also helpful for observation purposes, but should be reduced for image recording. The photomicrographs recorded with the light microscope may now be used to locate the same field, and matching scanning electron micrographs at the same magnification (and beyond) may be recorded. The matching photos help retain orientation. Conventional secondary electron images are adequate for revealing microstructure, while the backscatter mode provides atomic number imaging (Fig. 5-2) (Hall, 1981).

These same electron micrographs are used as maps indicating the X-ray analytical sites. Both bulk and point analyses may be conducted. The analyst must consider the effects of accelerating voltage, spot size, etc., on the origin of the X-ray signal. For example, during the point analysis of an inclusion it is very easy to excite the surrounding base metal as well as the inclusion, and misleading data can result (see Chapter 3). Similar analytical conditions apply to the analysis of in situ corrosion products and platings.

When examining the outermost layer of a polished specimen that is uncoated, orienting the sample with its edge parallel to the horizontal

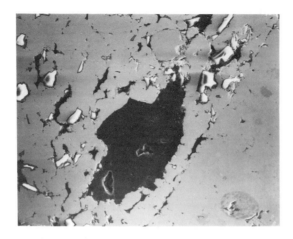

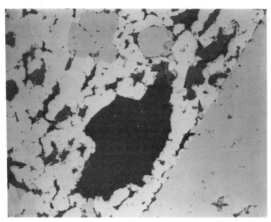

Fig. 5-2.   A polished sample viewed in the secondary (top) and backscattered (bottom) electron modes. (Courtesy of A. Laudate and JEOL)

axis of the CRT is helpful. Because charging is manifested in the horizontal direction, this artifact will appear above or below the region of interest. If the specimen edge intersects the vertical axis of the CRT, charging will appear across the CRT, thereby degrading image quality. This type of orientation is less important when the mount has been coated with a conductive thin film.

## REFERENCES

American Society for Metals (1972) *Atlas of Microstructures of Industrial Alloys*. Metals Handbook, 8th ed., vol 7. ASM, Metals Park, OH.
_____ (1973) *Metallography, Structures, and Phase Diagrams*, Metals Handbook, 8th ed., vol 8. ASM, Metals Park, OH.
Arrowsmith, A. W., et al (1978) Quantitative image analysis of toxic and brittle ceramics. *SEM, Inc.* 1:311.

Buehler, Ltd. (1981a) Metallographic sample preparation. Metal Digest 20(2):1. Buehler, Ltd., Lake Bluff, IL.

_____ (1981b) Metallographic sample preparation. Metal Digest 20(3):1.

Buhler, H. -E., and H. P. Hougardly (1980) *Atlas of Interference Layer Metallography*. American Society for Metals, Metals Park, OH.

Davidson, D. L. (1981) How to use the scanning electron microscope for failure analysis and metallography. *SEM, Inc.* 1:403.

*Dillinger, L. (1982) Automation in the metallography laboratory. *Amer. Lab.* 14(4):76.

*Gnizak, D. (1982) Light microscopes have key role in materials analysis. *Ind. Res. Dev.* 24(3):114.

Goodhew, P. J. (1973) Specimen preparation in materials science. In: *Practical Methods in Electron Microscopy*, vol 1, pt 2. (Glauert, A. M., ed.) American Elsevier, New York.

Hall, M. G. (1981) Metallography in the SEM. *SEM, Inc.* 1:409.

Johari, O., et al (1969) Microstructures of common metals and alloys as observed by the SEM. *IITRI/SEM*, p 279.

Kestel, B. J. (1981) *Polishing Methods for Metallic and Ceramic Transmission Electron Microscope Specimens*. ANL-80-120, Argonne National Lab., Argonne, IL.

Kotval, P. S. (1969) Comparative metallographic observations of surface shears due to a martensitic reaction using scanning electron microscopy, replication, and interferometry. *IITRI/SEM*, p 309.

*Leco Corp. (1977) *Metallography Principles and Procedures*. Leco Corp., St. Joseph, MI.

Liebl, I. (1981) Preparation of silicon chips. *Structure* 2:22.

Linke, U., and W. -U. Kopp (1981) Preparation of polished specimens and thin sections of plastics. *Structure* 2:9.

McCall, J. L., and P. M. French, eds. (1977) *Interpretive Techniques for Microstructural Analysis*. Plenum Press, New York.

_____ (1978) *Metallography in Failure Analysis*. Plenum Press, New York.

McCall, J. L., and W. M. Mueller (1974) *Metallographic Specimen Preparation: Optical and Electron Microscopy*. Plenum Press, New York.

*Nelson, J. A. (1976) Modern methods and materials for metallographic mounting. *Microstr. Sci.* 4:327.

Nelson, J. A., and E. D. Albrecht (1976a) The basics of metallography. *Heat Treating*, April Issue, p 2.

_____ (1976b) The basics of metallography. *Heat Treating*, June Issue, p 7.

Nelson, J. A., and R. M. Westrich (1974) Abrasive cutting in metallography. *Proc. 1973 Metallog. Symp.*, p 41. Plenum Press, New York.

Petzow, G. (1978) *Metallographic Etching*. American Society for Metals, Metals Park, OH.

Richardson, J. H. (1974) Specimen preparation methods for microstructural analysis. In: *Microstructural Analysis: Tools and Techniques*. (McCall, J. L., and W. M. Mueller, eds.) Plenum Press, New York.

Samuels, L. E. (1980) *Optical Microscopy of Carbon Steels*. American Society for Metals, Metals Park, OH.

_____ (1982) *Metallographic Polishing by Mechanical Methods*, 3d ed. American Society for Metals, Metals Park, OH.

Slepian, R. M., et al (1981) Semiautomatic metallographic preparation of silicon web samples. *Metallography* 14:213.

Stanton, R. W., and R. B. Finkelman (1979) Petrographic analysis of bituminous coal: Optical and SEM identification of constituents. *SEM, Inc.* 1:465.

*Struers, Inc. (1982) Struers Metallographic News: Special Issue on Sample Preparation. *Structure* 4:1.

Turner, J. F., and D. L. Rivenburgh (1979) Cleaning solution for metallographic mounts. *Metallography* 12:181.

Weidmann, E. (1982) Preparation of ceramic specimens for reflected light microscopy. *Metal Progress* 122(11):51.

---

*Recommended reading.

# 6

# Fracture Surfaces

## INTRODUCTION TO FRACTOGRAPHY

*Failure analyses* are conducted to determine how and why a component failed, and serve to define corrective actions to avoid future similar failures. Expert failure analysts follow a prescribed procedure during an investigation, the sequence of which may vary depending upon the nature of the failure. The American Society for Metals (1975c) outlines the stages of a failure analysis as follows:

1. Collection of background data and selection of samples.
   a. Determination of all details surrounding the failure.
   b. Determination of manufacturing, processing, and service history of the failed component.
   c. Preliminary reconstruction of the sequence of events leading to the failure.
   d. Photography of the failed component or structure at the site of failure.
   e. Samples selected for further analysis must be representative of the whole.
   f. If possible, similar components that did not fail should be selected for comparison.
2. Preliminary examination of the failed component.
   a. The specimens are visually examined with the unaided eye and a low-power stereo microscope prior to cleaning.
   b. The failed component is photographed and its dimensions are measured. If several pieces are involved, their relationships should be indicated.
3. Nondestructive testing.
   a. To detect surface cracks or discontinuities, magnetic particle inspection or liquid penetrant inspection may be conducted.

    b. To detect subsurface flaws, electromagnetic (eddy current) inspection, ultrasonic inspection, or radiography is necessary.

    c. Experimental stress analysis identifies the loads and stresses that can cause failure.

4. Mechanical testing.

    a. Hardness testing is used to compare the hardness of the failed component with that in the specifications to approximate the tensile strength of the material, and to detect work hardening or softening.

    b. Tensile testing also serves to relate the component's performance with the specified tensile properties.

5. Selection, identification, preservation, and cleaning of specimens.

    a. Fracture surfaces must be protected from mechanical damage or chemical attack to avoid destruction.

    b. Often, the fracture surface must be cleaned for identification of the mode of failure.

    c. If the specimen requires sectioning, careful sketches or photographs must be prepared showing the location of the section relative to the entire component.

    d. If the primary fracture surface is severely damaged, it may be necessary to open cracks and examine their surfaces.

6. Macroscopic examination and analysis of surfaces.

    a. Surfaces are examined with the unaided eye, then with a low-power stereo microscope to determine the configuration and general features of each fracture surface.

    b. Visual examinations provide a great deal of information, such as the stress system that induced failure, the origin of the fracture, and the direction of crack growth.

7. Microscopic examination and analysis.

    a. The fracture surface (or a replica) may be examined in the SEM and the mode of failure determined.

    b. The recording of stereo pairs assists in the interpretation of the relationships among features of the fracture surface.

8. Selection, preparation, examination, and interpretation of metallographic specimens.

    a. Metallography reveals the type and structure of the alloy, as well as undesirable surface effects (e.g., corrosion or work hardening) induced during manufacture or service.

    b. Specimens selected for metallography must be representative of the whole component.

9. Determination of failure mechanism: distinction between brittle and ductile fractures.

10. Chemical analysis

    a. Identification of the base metal composition (using, for example, atomic absorption spectroscopy) and comparison with specifications.

    b. Identification of corrosion products (e.g., X-ray spectroscopy, laser Raman spectroscopy, or X-ray diffraction).

11. Analysis of fracture mechanics.
    a. Measurement of fracture toughness.
    b. Evaluation of notch effects.

12. Testing under simulated service conditions. Although service failures are very difficult to reconstruct in the laboratory, data produced with proper interpretation of test limitations may be used to plan corrective action to avoid similar failures.

13. Analysis of all the evidence, formulation of conclusions, and report writing.
    a. The results of steps 1-12 are compiled and their interrelationships interpreted.
    b. Recommendations for corrective actions to avoid future similar failures are presented.

This very brief outline is a summary of the American Society for Metals (1975c) report. Interested readers are urged to consult that source for a comprehensive discussion of failure analysis. Similar perspectives are discussed by Colangelo and Heiser (1974), Boardman (1979), Tung et al (1981), and Wulpi (1985).

In this chapter, we are primarily concerned with *fractography,* defined as the study and documentation of fracture surfaces. Fractography is a subdivision of failure analysis: the purpose of fractography is to reveal the *mode of failure,* and when this information is combined with all other data, the failure analyst identifies the *cause of failure* (American Society for Metals, 1974c). The novice must appreciate that not all failures involve fracture: failures arise from hundreds of different mechanisms, and fractures are associated with only some failures. Described in this chapter are the general classifications and features of fractures. The novice is advised to consult Volume 10 of the Metals Handbook, Eighth Edition (ASM, 1975), for a more comprehensive survey of failure analysis.

Fractography relies upon visual interpretation of a fracture surface, with examinations beginning at the macroscopic level and proceeding to the microscopic level. Critical examination of a fracture surface with the unaided eye and with a low-power stereo microscope reveals extensive information. During a macroscopic examination the experienced analyst can define the fracture configuration and origin and identify the stress system that produced failure. Relative to the extant stress system, visual observations serve to classify a fracture mode as ductile or brittle. A *ductile failure* (or plastic failure) exhibits deformation (e.g., necking), which is induced when strains exceed roughly 5%. The absence of gross deformation indicates that fracture occurred at strains below 5%, and the fracture is classified as a *brittle failure* (nonplastic failure).

After classification of the fracture mode as brittle or ductile, the fracture surface is examined in the SEM to reveal the exact mode of failure

**Table 6-1.    Classification of Failure Modes**

| Ductile failures | Brittle failures |
|---|---|
| 1. Tensile overload<br>2. Shear overload<br>   a. Transverse shear<br>   b. Torsional shear<br>3. Bending overload | 1. Intergranular fracture mechanisms<br>   a. Stress-corrosion cracking<br>   b. Liquid-metal embrittlement<br>   c. Hydrogen embrittlement<br>   d. Creep<br>   e. Grain-boundary embrittlement<br>2. Transgranular fracture mechanisms<br>   a. Cleavage<br>   b. Stress-corrosion cracking<br>   c. Fatigue |

(Table 6-1). Ductile fractures are subdivided into tensile overload, transverse shear, torsional shear, and bending overload, all of which have different microscopic features. Brittle fractures are classified into intergranular or transgranular fracture mechanisms.

Intergranular fracture mechanisms are subdivided into stress-corrosion cracking, liquid-metal embrittlement, hydrogen embrittlement, creep, and grain-boundary embrittlement. Modes of transgranular fracture include cleavage, stress-corrosion cracking, and fatigue.

In some cases, the experienced fractographer will have identified the fracture mode prior to SEM examination, based upon information gleaned from the macroscopic examination and other stages of failure analysis. In such situations, the purpose of an SEM examination is to confirm the mode of failure. For example, many fatigue failures exhibit visible clamshell markings; fatigue can be confirmed in the SEM by imaging microscopic fatigue striations. Thus, the dividing line between macroscopic and microscopic analysis is not clearly defined: the successful analyst learns to rely upon a combination of information sources to define the mode of fracture.

The methods used in fractographic interpretation are discussed below. Also covered are the methods used to clean fracture surfaces, a very important stage in the analysis. Finally, it is assumed that the failed component can be directly examined in the SEM. If the specimen is too large to fit in the specimen chamber and cannot be sectioned, exact reproductions of the fracture surface, referred to as replicas, are prepared from selected sites and examined. Chapter 7 is devoted to replication, and Chapter 8 describes methods to render replicas electrically conductive. Although this chapter is devoted to metal fractures, note that similar fractures occur in ceramics and plastics. The brief bibliography following the references at the end of this chapter should be consulted for further information on fractography of these materials.

# METHODS OF FRACTOGRAPHY

As noted above, fractography is a subsection of failure analysis that reveals the mode of fracture. In the outline of the stages in failure analysis given on pages 95-97, fractography falls within steps 5-7. Each step is discussed below in greater detail. It is important to note that the sequence of preparation and examination is critical for the proper interpretation of fracture surfaces.

## Preliminary Examination

Fracture surfaces are fragile, subject to mechanical and environmental damage that can destroy microstructural features (Farrar, 1974). Consequently, fractured components must be carefully handled during all stages of an analysis. After photographic documentation, and for transfer from the site of failure to the laboratory, small components should be wrapped in a lint-free cloth or paper towel, placed in a waterproof container (e.g., sealable plastic bag), and labeled. Components should not be wrapped in cotton because the fibers tend to adhere to rough surfaces and are difficult to remove. Fracture surfaces which are too large to conveniently handle with this "wrap-and-bag" method are protected from oxidation by coating with fresh oil, axle grease, clear acrylic lacquer, or a montage of cellulose acetate replicas (Zipp, 1979). Under these circumstances, note that replicas alone preserve and entrap any surface debris, which often provides important clues to the failure mode or conditions during failure.

In the laboratory, the failed components are photographed and visually examined. If a component has fractured into several smaller pieces, their relationships should be determined and documented. During reconstruction, never actually fit fragments together because fine surface features will be destroyed. Also avoid touching the fracture surface; acids from your hands or fingers will corrode the metal. Reconstruction of a component fractured into many small fragments is analogous to putting together a three-dimensional jigsaw puzzle.

Photographs are important records used to document the progress of a fractographic analysis. They must be recorded before and after any surface treatment (e.g., cleaning) or sectioning. The adage that a photograph is worth a thousand words is valid for fractography. Refer to the American Society for Metals (1974d) for a thorough discussion of the photography of fracture surfaces.

During this stage, the analyst carefully measures and records the dimensions of the failed component and then subjects it to a thorough visual examination. This examination must be conducted before cleaning the specimen. Further, one must not assume that all material adhering to a fracture surface is foreign: debris and deposits can provide valuable information about a failure. For example, given a painted component, the presence of paint on a fracture origin is possible evidence of a crack in the component which existed prior to painting. Likewise, corrosion deposits are important clues to the mode of failure and aggressive species that may have caused or contributed to the failure. A visual appraisal by an observant analyst will differentiate fracture-related deposits from postfracture debris. In situations where the deposits are related to the failure, a cleaning method that preserves the deposits is necessary. Extraction replicas simultaneously clean the surfaces and preserve deposits. If the debris bears no relation to the fracture event, the surface may be cleaned without regard to preservation. In this case, the least aggressive cleaning method that exposes the native fracture surface is desired.

## Cleaning of Fracture Surfaces

A clean fracture surface is prerequisite for definition of the mode of failure. As with any other type of specimen, fracture surfaces must be clean for successful SEM imaging. Because fracture surfaces are fragile, their cleaning must be approached with caution and common sense. The cleaning methods are conveniently classified according to their degree of aggressiveness, as shown in Table 6-2. As a rule, the least aggressive method must be attempted before proceeding to more aggressive techniques, because the latter are capable of damaging the fracture surface. Dahlberg (1974, 1976), Zipp (1979), and Dahlberg and Zipp (1981) comprehensively review and compare each method, and their views are summarized below.

The least aggressive cleaning method is capable of removing loosely adhering dust or debris from the fracture surface. Short-haired *soft brushes* (e.g., a trimmed artist's paintbrush) or bursts of *compressed gas*

**Table 6-2.   Classification of Methods for Cleaning Fracture Surfaces According to Their Degree of Aggressiveness**

| Method | For removal of: | Degree of aggressiveness |
|---|---|---|
| Soft fiber brush and dry air | Loosely adhering debris and dust | Least aggressive |
| Organic solvents and ultrasonic bath | | ↓ |
| Toluene or xylene | Oil and grease | ↓ |
| Ketones | Varnish and gum | ↓ |
| Alcohol | Dyes and fatty acids | ↓ |
| Replica stripping | Insoluble debris and oxides | ↓ |
| Detergents (e.g., Alconox) | Corrosion products and oxides | ↓ |
| Cathodic cleaning | Deposits and oxides | ↓ |
| Corrosion-inhibited acids | Sulfides and oxides | ↓ |
| Acid etches | Oxides | Most aggressive |

are useful for removing dust. This method alone is rarely sufficient to clean a surface; more often, organic films (oil, grease, etc.) also obscure the surface. An ultrasonic treatment with an *organic solvent* followed by blowing with compressed gas removes both organic films and debris. Zipp (1979) recommends the various solvents listed in Table 6-2; for heavily contaminated surfaces, it may be necessary to pass through several changes or a series of solvents.

If the debris obscuring the surface is of interest, solvent cleaning should not be used immediately. For example, if the surface is corroded, it may be desirable to first analyze the composition of the corrosion products with energy-dispersive spectroscopy, then clean the specimen and examine the native fracture surface. Alternatively, the surface may be simultaneously cleaned and the reaction products preserved by preparing *extraction replicas* (synonymous with cleaning replicas). When a strip of cellulose acetate softened with acetone is placed over a fracture and firmly pressed into position (with one's thumb), the gel will encapsulate any material that is not part of the base metal. After the acetone has evaporated, the cellulose acetate retains the entrapped reaction products. The replica is then removed from the surface, and both the particles and their location relative to the fracture surface are preserved.

Heavily contaminated/oxidized surfaces may require sequential stripping of several replicas before the native surface is exposed. Those replicas applied first will contain heavier deposits than those applied subsequently, i.e., the first replica removes the most material (Fig. 6-1). The fractured component is then cleaned in an ultrasonic bath with acetone followed by blowing with compressed air, and the effectiveness of cleaning is evaluated with a stereo microscope.

Although extraction replicas are very effective for cleaning oxidized surfaces, their main disadvantage is that, despite solvent cleaning, fragments of cellulose acetate adhere to very rough surfaces. Because this material is nonconductive, it will charge during SEM irradiation and degrade image quality. This problem is aggravated if the replica is stripped before it has completely dried; the cellulose acetate must be dry. If particle preservation is not an issue, this problem is minimized by

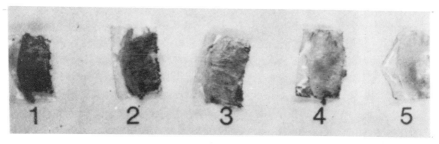

Fig. 6-1.   Replicas stripped sequentially from an oxidized fracture surface. Actual size. (Courtesy of Mr. Richard Zipp)

ultrasonic cleaning of the specimen with acetone between replicas. The final acetone rinse should be repeated two or three times with fresh acetone to ensure that all residual cellulose acetate has been removed. The details of replication, whether for surface reproduction or cleaning, are presented in Chapter 7.

Extraction replicas are effective for removing most oxidation products, but severely oxidized surfaces may require more rigorous cleaning. It must be understood that although cleaning will expose the metal surface beneath the oxide layer, oxidation itself has consumed some of the base metal, destroying the outermost layer of the native surface. Consequently the removal of oxide scale does not restore the fracture surface to its condition at the moment of failure. Under very harsh conditions, such as the environmental exposure of a failed component for months or years, the native fracture surface may be completely destroyed. Under less severe conditions, enough of the surface usually survives for definition of the fracture mode. Further, the very aggressive cleaning methods may themselves attack the base metal and erase all fine structural features. With these factors in mind, the analyst must be cautious and prepared to interrupt any of the following cleaning processes. One can readily resume the cleaning method, but the fracture surface cannot be restored once it has been obliterated by inappropriate cleaning methods.

Moderately aggressive cleaning with *water-based detergents* removes adherent oxides and corrosion products. Alconox is a popular detergent available from many laboratory suppliers. Zipp (1979) recommends a solution of 15 gm Alconox/350 ml water; the solution is heated to 90 °C and fractures are cleaned for 30 min. Simultaneous ultrasonic treatment is desirable. The specimen is then thoroughly rinsed with water followed by acetone and dried. Prolonging this treatment or increasing the concentration of the Alconox is ineffective and may cause attack of the base metal.

*Cathodic cleaning* is another moderately aggressive cleaning method for removal of oxides or heavy surface deposits. The specimen is made the cathode, an inert metal or graphite the anode, and both are submerged within an electrolytic bath of sodium cyanide, sodium carbonate, or sodium hydroxide (DeLeiris et al, 1966; Yuzawich and Hughes, 1978). Commercially available Endox 214 is another popular electrolyte (Dahlberg and Zipp, 1981). During the electrolytic reaction, the specimen is mechanically cleaned by the scrubbing action of hydrogen bubbles generated by the specimen. The specimen is rinsed in water and then acetone, dried, and examined. The cleaning should be periodically interrupted and its effectiveness evaluated with a stereo microscope; if required, cathodic cleaning may be repeated. High-strength steels or other alloys susceptible to hydrogen-induced cracking may be adversely affected by cathodic cleaning.

Sulfides and oxides are removed from fracture surfaces using *corrosion-inhibited acids*. The acid attacks and displaces the reaction products while the inhibitor protects the base metal from attack. However, the base metal will be attacked (etched) if the progression of cleaning is not carefully monitored. Ferrous alloys have been cleaned with 6N HCl containing 2 gm/l of hexamethylene tetramine (DeLeiris et al, 1966; Lane and Ellis, 1971; Dahlberg, 1974). Kayafas (1980) used 1,3-Di-n-butyl-2-thiourea to remove iron sulfide films. Both ferrous and nonferrous alloys may be cleaned with 2-butyne-1,4-diol inhibited HCl (Nathan, 1965; Farrar, 1974; Dahlberg, 1976). This corrosion-inhibited acid is prepared as follows:

| | |
|---|---|
| HCl (1.190 specific gravity) | 3 ml |
| 2-butyne-1,4-diol (35% aqueous) | 4 ml |
| Distilled water | 50 ml |

The specimen is cleaned by immersion in an ultrasonic bath for 30 sec, followed by rinsing with water and then acetone, and drying. Do not exceed a 30-sec exposure to these solutions; a prolonged treatment increases the probability of base-metal attack.

If the corrosion-inhibited acid method fails, as a last resort the specimen may be cleaned in a *weak acid or base*. This extremely aggressive method will attack the base metal unless constantly monitored. Knapp (1948) recommends weak acetic acid, phosphoric acid, or sodium hydroxide for cleaning ferrous alloys. Titanium alloys may be cleaned with nitric acid (Zipp, 1979). Aluminum alloys are cleaned with a mixture of orthophosphoric acid (70 ml of 85% aqueous solution), chromic acid (32 gm), and distilled water (130 ml), as described by Pittinato et al (1975). Following the brief submersion in an acid or base, the specimen is rinsed in water and then acetone, and dried. Water washing must be thorough to stop the reaction; residual acid or base will consume the base metal, and vapors will attack stereo microscope objective lenses.

To summarize, the analyst should always choose the least aggressive cleaning method that effectively exposes the fracture surface. The objectives of the study should be known before the specimen is cleaned; if debris removed from the surface requires preservation, extraction replicas should be prepared, because with other methods the surface materials are lost. One should also evaluate the effectiveness of the treatment while it is in progress; cleaning will not be adversely affected if stopped and restarted. Always evaluate cleaning with a low-power stereo microscope, and when the cleaning appears adequate, examine the specimen in the SEM. By following these precautions, damage of the surface of interest is avoided.

## Macroscopic and Microscopic Examinations

After the fracture surface has been cleaned, another series of photographs are recorded. This set serves to document both the appearance of the fracture and the progress of the analysis. There are clear advantages to recording both conventional photographs and SEM micrographs in fractography. Macrophotographs are used to map the fracture surface and to identify sites selected for SEM. This permits direct correlation between different levels of magnification. Further, some fracture surfaces exhibit different textures or colors across their diameter that are readily visible; SEM images should be recorded in these different zones and their location indicated on a macrophotograph. For example, fatigue fractures initiate, propagate, and finally undergo fast fracture: each of these stages appears different from the others. A macrophotograph serves as a map to show the location of SEM micrographs of each stage. Orientation is also preserved, which is important when the fracture origin is not clearly defined at the macroscopic level. Finally, SEM micrographs usually encompass only a small area relative to the entire fracture surface. By recording a series of micrographs from low to high magnification, the orientation of the fracture surface and progression of the analysis are uninterrupted.

During this phase of the fractographic analysis, the surface is again thoroughly examined with the naked eye and a low-power stereo microscope. In many cases the experienced analyst can identify the stress system operative during failure and the origin and direction of propagation of the crack front (American Society for Metals, 1974a and 1974b; Anderson, 1983). Several macroscopic observations define these parameters. First, the analyst classifies the failure mode as brittle or ductile, based upon, respectively, the absence or presence of gross deformation of the component. If little deformation is visible, the stress system imposed less than roughly 5% strain on the component during failure. In comparison, the presence of necking or distortion implies that strains exceeding 5% existed during failure. Recall, however, that the macroscopic manifestation of creep does not follow this rule.

Other macroscopic features identify the fracture origin and thus the direction of crack propagation. *Radial marks* are observed in both ductile and brittle failure modes and are indicative of unstable crack growth (Fig. 6-2). The term radial mark implies that the specimen was rod-shaped; e.g., tensile fractures exhibit radial marks that fan out from the origin. In specimens that fractured at high velocity, these marks are referred to as *chevrons* (synonym: herringbone pattern), which are linear indications that point back to the crack-initiation zone (Fig. 6-3). Chevrons are commonly seen in brittle fracture of low-carbon steel plate. Room-temperature fractures of high-strength steels (tempered martensite) often exhibit fine radial marks or chevrons, whereas medium-strength steels (having martensitic structures) have coarse radial marks or chevrons. At lower temperatures, both medium- and high-strength steels exhibit fine radial marks or chevrons.

Indications of progressive propagation of a crack are referred to as *arrest marks*, with each arrest mark corresponding to an overload. Arrest marks may be observed in progressive fracture modes such as stress-

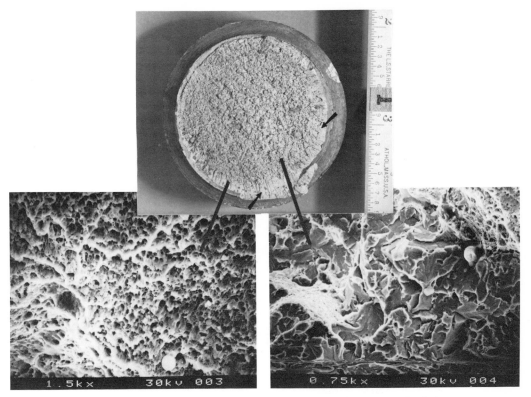

Fig. 6-2.  Radial marks (arrows) in the fibrous zone of a bolt
fractured under conditions of tensile overload. The morphologies
of the different texture zones are shown in the SEM micrographs:
at left, ductile fracture; at right, transgranular fracture.

corrosion cracking (Fig. 6-4) and fatigue. In fatigue, arrest marks are
referred to as *clamshell markings* (synonymous with beachmarks and
conchoidal marks), and radiate away from the origin in concentric semicircles (Fig. 6-5). The clamshell markings can be resolved into minutely
spaced fatigue striations in the SEM, whereas the arrest marks sometimes visible with stress-corrosion cracking are simply regions of overload fracture that alternate with the corrosion event.

*Ratchet marks* may be observed in either brittle or ductile fracture
modes, and are manifested as small steps on the fracture surface that
result from the junction of separate crack fronts propagating along different planes. Ratchet marks are often visible on components which fractured by low-cycle, high-stress fatigue.

Using any of these indications as guidelines, the origin of the fracture
may be confirmed with the SEM. If the macroscopic indications are not
clearly defined, probable origins can be identified and confirmed or
rejected during the SEM analysis. The identification of the direction of
crack growth is extremely important in fractography: the location of
the crack origin can bear directly on the cause of failure.

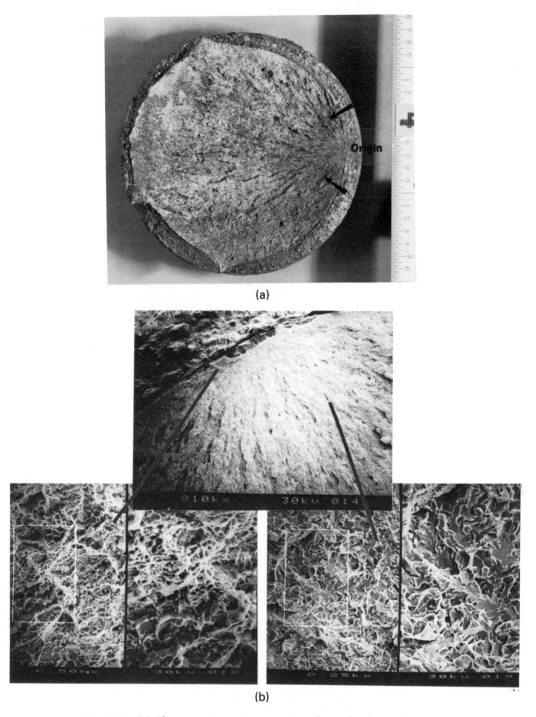

(a)

(b)

Fig. 6-3. (a) Chevrons (arrows) emanating from the fracture
origin in a bolt that failed under conditions of bending overload.
(b) SEM micrographs of the origin and fracture surface shown in
(a).

Fig. 6-4. Stress-corrosion cracking arrest marks arising from differences in the rate of penetration of corrosion on the fracture surface. 4×. (From *Metals Handbook*, 8th ed., vol 9, p 31. Courtesy of American Society for Metals)

Fig. 6-5. Macroscopic clamshell markings (arrows) radiating away from the origin in a fatigue failure.

Once the origin has been identified, its position is indicated on the macrophotograph, and the same area is located in the SEM. A micrograph at very low magnification (10-20 diameters) serves as a good transition between the macrophotograph and higher-magnification micrographs. If possible, the entire area of the fracture surface should be recorded at low magnification; if the sample is too large, it may be necessary to prepare a montage of overlapping micrographs. Higher-magnification micrographs are then recorded. Again, a progression from low to moderate to high magnifications helps retain perspective and assist in interpretation. The $X$ and $Y$ specimen movements are of course variable during an analysis, but rotate and tilt should be maintained constant throughout the SEM examination. If the orientation is altered by rotating the specimen, another low-magnification micrograph should be recorded. Images recorded at different orientations are difficult to interpret, particularly when attempting to identify the fracture origin. As will be discussed, the orientation of dimples in a ductile rupture is used to locate the origin; if the analyst does not maintain the same orientation throughout the SEM analysis, almost useless information will result.

Additional series of micrographs are then recorded in other preselected areas of the fracture surface. Different areas may include those having different macroscopic surface textures or colors, as with the three fatigue zones described earlier. The required range of magnifications will vary according to the mode of fracture. For example, magnifications up to several thousand diameters are often required to characterize dimples and their orientation in ductile rupture, whereas in many intergranular fractures magnifications less than 500× may be sufficient to define the fracture mode. The novice will develop a "feel" for the most meaningful magnification levels for a given fracture type, and will learn at what range empty magnification is reached. The following examples of ductile and brittle fracture modes provide general indications of desired magnification levels. The novice must appreciate that the information presented below is necessarily in very general terms. A full description of any one of these fracture modes would constitute a separate textbook. Therefore, the analyst must not make snap judgments on the basis of the general information presented in this text but is challenged to interpret fracture surfaces in the context of the entire failure analysis.

Several excellent reviews of fractography are available, namely Beachem (1968a and 1968b), American Society for Metals (1974a-1974c), Broek (1974), Pelloux (1974 and 1976), Strauss and Cullen (1978), and Tung et al (1981). Atlases of fractographs are also available (Phillips et al, 1965; American Society for Metals, 1974; Pittinato et al, 1975; Bhattacharyya et al, 1979; and Engel and Klingele, 1981). These atlases cover a broad spectrum of alloys which were subjected to various testing conditions, and are recommended sources for both the novice and the experienced failure analyst.

# DUCTILE FRACTURE MODES

Crack extension in ductile fractures involves plastic deformation and is often observed under conditions of tensile overload or shear. Ductile fractures appear dull and nonreflective to the naked eye; with the SEM, those fractures exhibit unique structures referred to as *ductile dimples*. Dimples are hemispheroidal cavities, frequently containing an inclusion or precipitate. The dimples form by microvoid nucleation at these inclusion sites (where local plastic deformation is high), and under increasing strain the microvoids grow, meet one another (coalesce), and eventually separate (rupture), thereby producing the dimpled appearance (Beachem and Meyn, 1968). Therefore, dimples are one-half of a microvoid. The mechanism of ductile fracture is often referred to as *microvoid coalescence*, which describes the mode of crack growth.

SEM examination and interpretation of microvoids exhibited by ductile fracture surfaces reveal information about the type of loading experienced during fracture, the direction of crack progression, and the relative ductility of the material (Beachem, 1963 and 1975; Beachem and Yoder, 1973). First, the shape of the dimples is determined by the type of loading the component experienced during fracture, and the orientation of the dimples reveals the direction of crack extension. *Equiaxed dimples* are cup-shaped and form under conditions of uniform plastic strain in the direction of applied stress; equiaxed dimples are typically produced under tensile overload conditions. In comparison, *elongated dimples* shaped like parabolas result from nonuniform plastic strain conditions such as bending or shear overloads. These dimples are elongated in the direction of crack extension and therefore reveal the fracture origin. The shape and orientation of microvoids as related to conditions of loading are discussed below.

The size of the microvoids is related to the ductility of the specimen, the size and number of inclusions in the specimen, and the relative level of stress during fracture. In general, ductile materials (e.g., pure copper) exhibit larger dimples than harder materials (e.g., carbon steel) fractured under the same conditions. The sustained application of a low stress level (e.g., progressive overload) produces larger microvoids than does a sudden overload. The number of inclusions or precipitates is another factor influencing the size and number of microvoids (Gurland and Platean, 1963; Beachem and Pelloux, 1964). Edwards (1964) observed that as temperature increased, the microvoid depth also increased. All of these factors contribute to a better understanding of ductile fracture modes.

**Tensile overload.** Specimens subjected to uniform plastic strain conditions such as axial tensile overload exhibit equiaxed dimples (Fig. 6-6). The dimples are cup-shaped and point "upward," indicating that the direction of applied stress is axial. The dimples observed on mating fracture surfaces of tensile-overload specimens exhibit the same shape

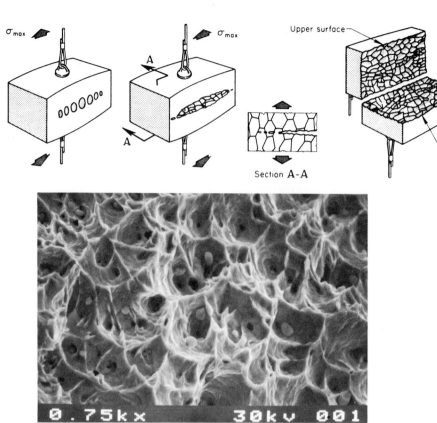

Fig. 6-6.    Formation of dimples under conditions of tension using a copper test specimen. Note that the dimples are equiaxed. (Sketch courtesy of American Society for Metals)

and orientation. This morphology is typical of tensile-test specimens, but is rarely encountered in service failures because it is more difficult to fracture a component in pure tension than it is by shear or bending. The reader can confirm this with a paperclip. It is very difficult or impossible to fracture the paperclip (by hand) in pure tension, but it readily fractures if a shear or bending moment is introduced.

**Transverse shear overload.** Under plane-stress conditions, a specimen in tension will fracture at a 45° angle relative to the specimen thickness. Shear dimples shaped like parabolas are observed on the fracture surface, and they are elongated in the direction of crack extension (which in turn corresponds to the direction of shear force). The stress conditions imposed during transverse shear predict that the dimples on mating fracture surfaces point in opposite directions, which correlates with the shear moment (Fig. 6-7). Components that have failed under transverse shear overload conditions exhibit a 45° fracture surface with a shear lip at the final fracture zone. The SEM serves to confirm ductile fracture and the direction of crack progression; positive

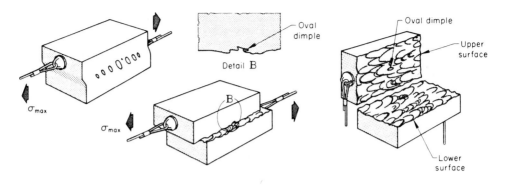

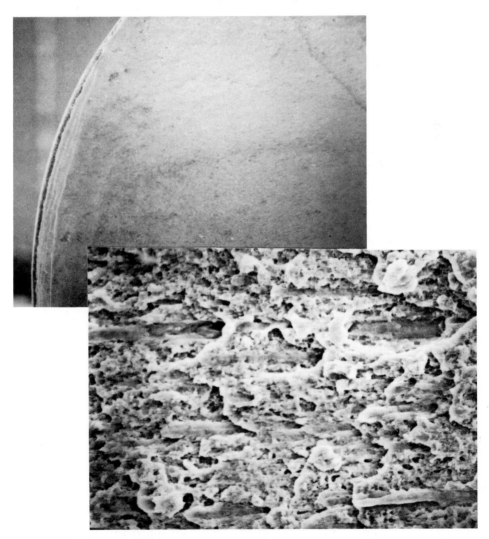

Fig. 6-7. Fracture of a high-strength steel under conditions of transverse shear overload. Micrographs at 25× (top) and 1000× (bottom). (Sketch courtesy of American Society for Metals)

identification of transverse shear overload as the effective stress system is made by examining both mating fracture surfaces. Because specimen orientation is critical, low- to high-magnification micrographs should be recorded in several areas of each fracture surface.

**Torsional shear overload.** Under conditions of torsional shear overload, a component is subjected to mutually perpendicular tensile and compressive stresses at a 45° angle relative to the specimen axis. The shear stresses are induced in the longitudinal and transverse directions, and the fracture surface exhibits a swirled or rotary deformation. In pure torsional shear overloads the final fracture zone is centered; if bending was introduced during stress, the final fracture zone will be offset from center (Fig. 6-8). The twisting observed at the macroscopic level is repeated at the fine structural level; in the SEM, elongated dimples which change in direction (rotate) around the circumference of the fracture surface are visible. The purpose of an SEM examination of components fractured under these conditions is to define the direction of crack extension. The importance of maintaining specimen orientation from low to high magnification and recording several micrographs around the circumference of the fracture surface is clear.

**Bending overload.** Components that have fractured by ductile bending overload exhibit the effects of both compression and tension: the

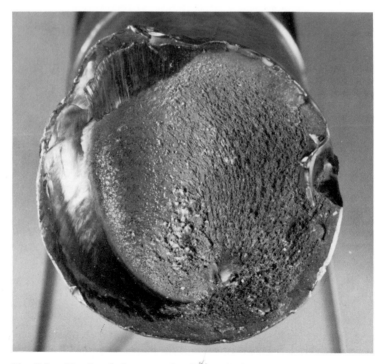

Fig. 6-8.   Torsional shear overload of a shaft. Note the rotary deformation of the fracture surface.

convex surface is held in tension, and simultaneously the concave surface is compressed (Fig. 6-9). SEM reveals that dimples are elongated in the direction of crack progression and exhibit the same orientation on mating surfaces. Bending overload can often be identified macroscopically; SEM serves to confirm the ductility of the fracture.

**Examination of ductile fracture surfaces.** Magnifications of roughly 3000-5000× are necessary to resolve fine microvoids. To make an easy transition from low to high magnifications, the microscopist should record a series of micrographs at progressively higher magnifications, then stop at the level where the shape and orientation of the dimples are clearly represented. At the moderate and high levels of magnification (roughly 500-5000×), stereo pairs should be recorded if the orientation of the dimples is unclear. Recall from Chapter 2 that spatial distortions are introduced by tilting the specimen; when assigning the direction of crack propagation on the basis of microvoid orientation, the microscopist should be very careful to take into account the beam-specimen-detector geometry. Under these conditions, a stereo pair will provide a more faithful rendition of the surface features than will a single micrograph.

A typical sequence of micrographs taken at roughly the following magnifications provides a great deal of information about ductile fractures: 10× (to relate the SEM appearance to the macrophotograph), 100× (a typical field of view), 500× (interrelationships among dimples), and 1000-5000× (shape and size of microvoids). Do not rotate or tilt the

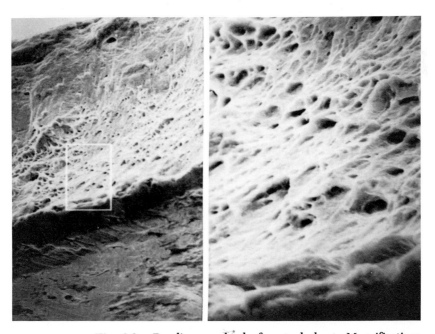

Fig. 6-9. Bending overload of a steel sheet. Magnifications: 250× (left); 1000× (right).

specimen between recordings; a change in sample orientation will produce confusing information. If tilt or rotate is changed, indicate the new orientation on the low-magnification micrograph.

Additional information about the specimen is provided by a qualitative X-ray analysis of the base metal and inclusions. Because the inclusions frequently are located deep within the microvoid, it may be difficult to obtain an X-ray signal solely from the inclusion because of absorption by the surrounding dimple. Further, the X-ray excitation volume may encompass both the inclusion and the underlying base metal. Usually manipulating tilt and rotate will produce an adequate count rate. A spectrum of the base metal may then be stripped from that of the inclusion (plus base metal) to reveal more valid compositional information about the inclusion alone.

If a more rigorous analysis of the inclusions is required, extraction replicas (prepared exactly as in the cleaning method described earlier) may be stripped from the fracture surface. Loosely held inclusions will be entrapped by the replica, which is then carbon-coated and examined in the microscope. Because the mean atomic weight of cellulose acetate is very low, it will not significantly affect the spectrum. This may be confirmed by recording an X-ray spectrum of the replica alone, and stripping this spectrum from that of the particle.

## BRITTLE FRACTURE MODES

Failed components which exhibit little or no deformation, i.e., have failed in a nonplastic mode, are classified as *brittle fractures*. The lack of deformation implies that fracture occurred at strains below 5%.

Brittle fractures are subclassified into intergranular and transgranular fractures. In an *intergranular fracture,* the crack propagates between the grains of the metal and produces a rock-candy or faceted appearance. Intergranular fractures occur in stress-corrosion cracking, hydrogen embrittlement, and creep. *Transgranular fractures* propagate through the metal grains and present a flat, bright appearance. Types of transgranular fracture include cleavage, fatigue, and other modes of stress-corrosion cracking. Each of these fracture modes is discussed below.

### Intergranular Fracture Modes

During *intergranular fracture* (synonymous with grain-boundary separation) the crack follows a path along grain boundaries. This mode of fracture is manifested as a rock-candy or faceted appearance. Intergranular fractures may arise when there are marked differences between the properties of the grains and the matrix between the grains, or when environmental conditions promote attack of the intergranular matrix.

*Grain-boundary embrittlement* arises from the thermally activated segregation of impurities along grain boundaries or the depletion of the intergranular matrix. The differences between the properties of the grains and the intergranular matrix promote separation along the grain boundaries. Another elevated temperature failure mode is *creep,* which can occur in steels and superalloys under conditions of simultaneous high temperature and sustained tensile stress. In creep, crystallographic re-alignment of grains produces vacancies at grain boundaries, and as the vacancies enlarge and join, intergranular separation results.

Adverse environmental conditions combined with low but sustained tensile stresses also promote intergranular fracture. *Stress-corrosion cracking\** is caused by corrosive attack of the grain-boundary matrix. *Liquid-metal embrittlement* occurs when a grain boundary is attacked by a liquid or molten metal.

It is important to understand that it is the synergistic effect of adverse conditions and sustained stress that promotes intergranular fracture. Therefore, a valid interpretation of such fractures is heavily dependent upon learning the manufacture, processing, and service history of the failed component. Although it is easy to recognize intergranular fracture, identification of the cause of fracture is more complex. Each of the fail-ure modes leading to intergranular fracture is discussed below.

**Grain-boundary embrittlement.** The segregation of thermally acti-vated impurities at grain boundaries promotes intergranular fracture along this weakened path (American Society for Metals, 1974c). This is often associated with low-alloy steels subjected to high temperature, which induces temper embrittlement and segregation or leaching at grain boundaries. For example, if a low-alloy steel component is heated to its embrittlement temperature (~700 °C), low-melting-point sulfides can segregate at grain boundaries. If the component is cyclically cooled and heated, the sulfide phase can melt or weaken. Under conditions of low stress, shock loading, or impact, the component fractures along the path of least resistance, which is the weakened grain boundary (Fig. 6-10). Similar causes of intergranular fracture are depletion at grain boundaries and the formation of a precipitate or a continuous film at grain boundaries.

**Intergranular stress-corrosion cracking (SCC).** Stress-corrosion cracking arises when a component is subjected to a sustained tensile stress while in a corrosive environment. The maximum tensile strength is typically much lower than the normal yield strength of the material, but that strength is severely undermined by certain environments. The combined effects of relatively mild corrosion and stress are additive, and the component fails much more quickly than if subjected to corrosion or stress alone (American Society for Metals, 1974e and 1974f).

High-strength aluminum alloys under atmospheric corrosion condi-tions can exhibit SCC: intergranular fracture in these alloys is uniquely

---

*Stress-corrosion cracking may also produce transgranular fracture, which is discussed later in this chapter.

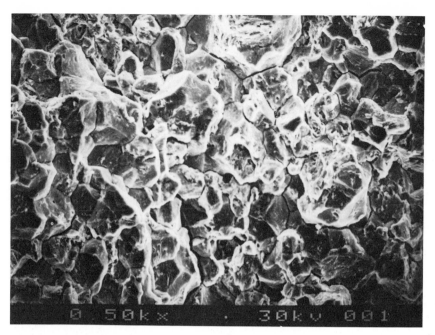

Fig. 6-10.   Intergranular fracture of an embrittled low-alloy steel.

caused by SCC (Fig. 6-11). Cuprous alloys, especially cartridge brass, are susceptible to SCC in ammoniacal ($NH_4^-$) environments. Although these fractures are usually intergranular, some transgranular fracture may also be seen. In low-carbon structural steels, intergranular SCC can occur in the presence of hot concentrated nitrate or caustic solutions. Both mild steels and stainless steels are subject to SCC in the presence of solutions containing the hydroxyl ion; fracture can occur by either intergranular or transgranular fracture, or both. High-strength steels are also susceptible to intergranular SCC in water. Titanium alloys in a high-temperature environment of sodium chloride fracture intergranularly, although regions of transgranular fracture may be observed (Phillips et al, 1965).

Intergranular fracture surfaces caused by stress-corrosion cracking can exhibit macroscopic *progression marks* which correspond to overloads (Fig. 6-4). These progression or arrest marks can be confused with the macroscopic clamshell markings seen in fatigue failures. With the SEM, however, the clamshell markings are resolvable into numerous fine fatigue striations, which are not present in SCC arrest marks. *Corrosion fatigue* is another mechanism of failure, but this is often a transgranular fracture mode usually initiated at a pit or crack (American Society for Metals, 1975a; Liaw et al, 1982). Determining the primary mechanism of failure in corrosion fatigue is sometimes difficult, although these cracks usually branch during propagation whereas "pure" fatigue frac-

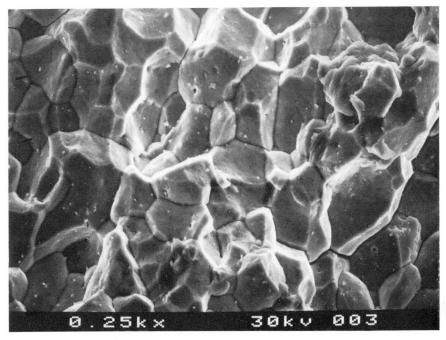

**Fig. 6-11.** Grain-boundary separation induced by atmospheric SCC of a high-strength aluminum alloy.

tures rarely branch. This branching in corrosion fatigue can be mistaken for stress-corrosion cracking.

**Liquid-metal embrittlement.** This failure mode is caused by the attack of grain boundaries by a liquid metal (American Society for Metals, 1975e). Liquid-metal embrittlement is often categorized as a special form of stress-corrosion cracking, although penetration by the liquid metal can occur with or without stress. Because stress is essential to SCC but not to liquid-metal embrittlement, these failure mechanisms are discussed separately.

Aluminum and copper alloys are subject to intergranular attack by mercury, and steels are attacked by molten tin or cadmium. In intergranular failures of steel plated with a low-melting-point coating, liquid-metal embrittlement may be suspected if the component has been exposed to high temperatures. The embrittling constituent is readily identified by energy-dispersive spectroscopy of the fracture surface.

**Hydrogen embrittlement (HE).** Hydrogen embrittlement is caused by the diffusion of hydrogen atoms into metal from high temperature, high pressure gas, corrosion, welding, or plating operations. When combined with residual or applied stress, the fracture is usually intergranular. In heat-treated high-strength steels, hydrogen dissolved in the alloy causes hairline cracking and loss of ductility. The pathway of the hairline cracks follows prior austenitic grain boundaries, and the cracks are induced when the damaging effect of dissolved hydrogen is super-imposed on the stresses that accompany the martensite-to-austenite transition (American Society for Metals, 1974c and 1975d; Bandyo-padhyay and McMahon, 1983; Brooks et al, 1983; Kwon and Kim, 1983; Landon et al, 1983). If initiating cracks are not pre-existent, they can initiate beneath the surface of the component where triaxial stress is highest.

Hydrogen embrittlement can combine with fatigue and produce a mixed-mode type of fracture. Schuster and Altstetter (1983) have observed this phenomenon in stainless steels, and Fig. 6-12 shows this effect in a 2014-T6 aluminum casting. In this case, a fatigue crack initiated and its progress accelerated through tiny pockets of hydrogen. Note the presence of a small nucleating particle within the pocket, and the surrounding zone of ductility.

In quenched-and-tempered high-strength steels, the distinction between HE and SCC may be difficult, because both failure modes are manifested as intergranular fractures. However, in this context, HE is the dominant mode of failure for high-strength steels (American Society for Metals, 1974c). In addition to determining the manufacture, processing, and service history of the component, several features may be used to distinguish intergranular SCC and HE. If hydrogen damage originates from plating operations, there is no evidence of corrosion, and grain

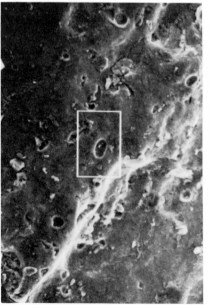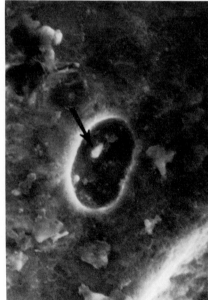

Fig. 6-12.   Fatigue failure of an aluminum casting accelerated by the presence of microscopic pockets of hydrogen. Note the presence of a nucleating particle (arrow). Magnifications: 300× (left); 3000× (right).

boundaries are very smooth. A study of SCC in high-strength aluminum alloys by Pickens et al (1983) indicated not only that HE is closely linked with SCC but that it may be the primary cause of failure under certain conditions. Corrosion-induced hydrogen damage is sometimes suggested by the corroded appearance of the fracture surface or the outer diameter of the component adjacent to the fracture. At the microscopic level, Phillips et al (1965) have observed a larger number of tear ridges in HE than in SCC intergranular fractures, and more secondary cracking in SCC than in HE failure modes. If a fracture surface has suffered corrosion or oxidation after the fracture event (e.g., a service failure exposed to the environment for several months), the analyst cannot distinguish SCC from HE.

**Creep.** Creep is a form of strain induced under high temperature and constant load or stress conditions, which cumulatively reduce the mechanical strength of a material. When a component has reached the limit of its mechanical strength, it fractures by intergranular *stress rupture*, also referred to as creep rupture.

The mechanism of creep is divided into three consecutive stages (American Society for Metals, 1975g). Primary creep occurs during the initial stressing of the component, and refers to the thermal activation of plastic strain conditions which induces realignment of the crystals through grain-boundary sliding. Consequently, primary creep causes a change in the material by a deformation or distortion mechanism. Secondary or steady-state creep is a transitional stage representing a period of alternate work hardening and recovery. This stage evolves into tertiary creep, where an increasing rate of extension causes wedge-type cracks to form at grain-boundary triple points. Voids are created and grain-boundary porosity develops. The grain-boundary porosity limits the mechanical strength of the component, which then fractures along the weakened grain boundaries. The fractures in creep rupture are predominantly intergranular, although some transgranular cleavage may be observed. The zones of transgranular fracture probably initiate at intergranular fissures that have decreased the available cross-sectional area and raised the stress (American Society for Metals, 1975g). A more comprehensive discussion of the mechanism of creep is available in Metallurgical Transactions (1983); Sadananda and Shahinian (1983) have evaluated the creep behavior of several high-temperature alloys.

The macroscopic appearance of creep-induced failures is a function of the rate of creep. *Short-term creep* failures arise if temperature or stress suddenly increases beyond tolerance levels, and they exhibit elongation or necking similar to that in ductile tensile failures (Fig. 6-13a). Despite their macroscopic appearance, ruptures arising from short-term creep are intergranular. *Long-term creep* failures appear brittle, i.e., without necking or distortion. Therefore, visual examination is all that is necessary to determine whether short-term (ductile appearance) or long-term

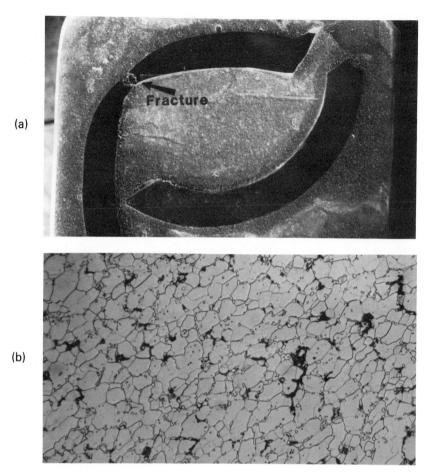

**Fig. 6-13.**    Creep-induced failure of a boiler plate. (a) A polished cross section of the plate (at about 2 ×) exhibits necking, a feature of short-term creep. (b) Intergranular voids (dark areas) in an area near the fracture surface.

(brittle appearance) creep was the mechanism of failure. Creep-rupture failures are confirmed by metallography; intergranular voids or cavities are visible on either side of the fracture surface (Fig. 6-13b). Although SEM may be desired for characterization of the fracture surface, it is not required for the majority of creep-related failures. As with other failure modes, the analyst must identify the service history of the component. By definition, the identification of creep presupposes that the component was subjected simultaneously to elevated temperatures and sustained stress.

**Examination of intergranular fracture surfaces.** The characterization of intergranular fractures by SEM proceeds from low to moderate magnification levels; features of individual grain facets may be recorded at higher magnifications. It is extremely important that several micrographs be recorded to demonstrate the relationships among grains; i.e., the rock-candy appearance of intergranular fracture must be demonstrated. The desired magnification level will vary among alloys but is generally in the 100-750× range. Recording only higher-magnification micrographs does not give a true picture of the fracture surface morphology, because the area of a given field of view is proportionally smaller as magnification is increased. For example, microvoid coalescence may be observed at grain interfaces under conditions where the yield strength of the grain-boundary zone is less than that of the matrix (Fig. 6-14). Grain facets may also exhibit small tear ridges (Beachem, 1972). If the microscopist recorded only these regions, misleading data would result. Rather, the microscopist should record several micrographs at increasing magnification levels for a more accurate portrayal of the fracture surface. Note that when areas of ductility are visible on a fracture surface that is predominantly intergranular, one concludes that the primary fracture mode was indeed intergranular.

As mentioned above, another purpose of SEM is to differentiate SCC progression marks from fatigue clamshell markings. The SCC arrest marks are not resolvable into closely spaced striations, which would confirm fatigue. Also in SCC, the corrosive agent may be identified with energy-dispersive spectroscopy, although Auger analysis may be required in cases involving alloy segregation (e.g., Schmerling et al, 1981).

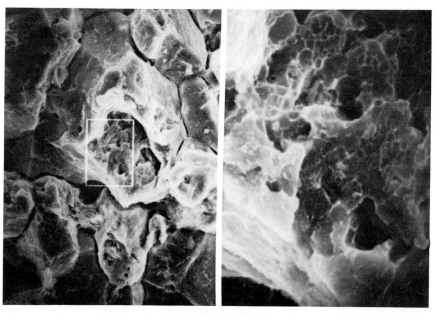

Fig. 6-14.   Ductile dimples observed on grain facets in an intergranular fracture. Micrograph at right (2500×) is very misleading; lower-magnification micrograph at left (250×) better characterizes the fracture surface.

## Transgranular Fracture Modes

Transgranular fracture modes are a type of brittle fracture characterized by the propagation of a crack through the grains or crystals of the material. The modes of failure that induce transgranular fracture are cleavage, stress-corrosion cracking, and fatigue.

**Cleavage.** Cleavage is a fracture mode associated with body-centered cubic (BCC) crystals (e.g., tungsten, molybdenum, and chromium) and hexagonal close-packed (HCP) crystals (e.g., zinc and magnesium). In the BCC and HCP crystal systems, as well as in ferrous and low-carbon steels, cleavage is manifested under conditions of high triaxial stress, at high deformation rates (e.g., overload and impact), and under low-temperature conditions (Nielsen, 1968; American Society for Metals, 1975c). The face-centered cubic system, which includes aluminum and its alloys, only rarely exhibits transgranular cleavage.

Cleavage is heavily dependent upon crystal orientation: i.e., the fracture pathway is defined by crystallography. In the BCC system, cleavage usually occurs on the {100} plane, whereas in the HCP system cleavage occurs on the {0001} basal planes. Beachem (1968b) gives a more comprehensive description of the mechanism of crack propagation in these systems.

A visual examination of a cleavage fracture reveals brightly reflecting facets, which in the SEM appear as very flat surfaces (Fig. 6-15). At

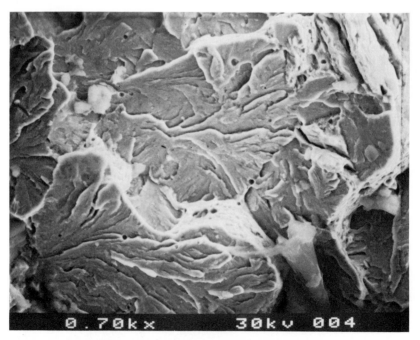

**Fig. 6-15.** Cleavage in a low-carbon steel specimen impact fractured at liquid nitrogen temperatures.

higher magnifications, the facets reveal features related to the direction of local crack propagation, which can in turn be related to the origin of the primary crack (Burghard and Stoloff, 1968; American Society for Metals, 1974c). *River patterns* (or river marks) represent steps between different local cleavage facets at slightly different heights but along the same general cleavage plane. The river patterns arise from fracture along second-order cleavage planes, and are so named because they join like river tributaries in the direction of local crack propagation. This minimizes the energy of fracture. When defining the direction of crack propagation, be aware of the fact that inclusions or precipitates can modify the local direction of propagation. The overall direction of crack propagation may be assigned only after confirming the orientation of the river patterns in several areas on the fracture surface.

Cleavage facets in low-carbon steels and iron can also display *tongues*, which are very fine slivers of metal (Fig. 6-16). They are usually aligned along defined crystallographic directions and result from cleavage across microtwins formed by plastic deformation at the tip of the primary crack. Localized zones of microvoid coalescence or intergranular fracture may also be observed.

As with other fracture modes, it is important to record micrographs at a range of magnifications, particularly if the crack origin is to be located. The overall direction of propagation should be defined by examining

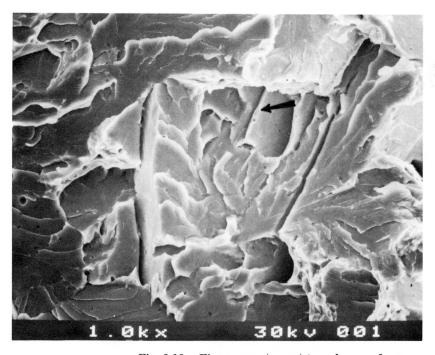

Fig. 6-16.   Fine tongue (arrow) in a cleavage fracture.

several areas across the diameter of the fracture surface at moderate (500-1000×) magnifications. The location and directionality of secondary cracks, if present, should also be documented.

**Transgranular stress-corrosion cracking.** In certain alloys, simultaneous exposure to an aggressive environment and sustained tensile stress can induce transgranular stress-corrosion cracking. The alloys most susceptible to transgranular SCC are austenitic stainless steels (e.g., 18Cr-8Ni type) and magnesium. Austenitic stainless steels exposed to aqueous environments containing chloride ions are susceptible to highly branched transgranular SCC (American Society for Metals, 1974c; Andersen and Duquette, 1980; Bandyopadhyay and Briant, 1983). One such failure is shown in Fig. 6-17.

Fractures caused by stress-corrosion cracking can be manifested as intergranular or transgranular fractures depending upon variables such as the state of cold work or heat treatment of the alloy. In general, low-carbon steels and aluminum alloys subjected to SCC exhibit intergranular fracture, while austenitic stainless steels and magnesium alloys exhibit transgranular fracture. Other alloys can exhibit both transgranular and intergranular fractures; for example, stainless steels exposed to hydroxyl ions can display a combination of the fracture modes. Fractures of mild steels in hydroxyl ion environments are usually intergranular with some transgranular features. Likewise, fractures of brass in ammoniacal environments are usually intergranular with minor trans-

Fig. 6-17.   Polished section of an austenitic stainless steel specimen exposed to chloride ions. Failure occurred by transgranular SCC. Note the extensive branching of the multiple cracks.

granular features. When these mixed-mode fractures are encountered, the analyst usually assigns the predominant fracture mode.

Fractures induced by transgranular SCC are examined in the same manner as cleavage fractures. Corrosion deposits may be analyzed with energy-dispersive spectroscopy, although Auger analysis may be required for positive identification. Valid interpretations require correlation of all factors surrounding the failure.

**Fatigue.** *Fatigue* is defined as the progressive localized permanent structural change that occurs in a material subjected to repeated or fluctuating strains at stresses having a maximum value less than the strength of the material (American Society for Metals, 1975b). Fatigue is caused by the simultaneous actions of cyclic stress, tensile stress, and plastic strain: cyclic stress initiates the fatigue crack, and tensile stress causes crack propagation. Among all causes, fatigue is reputed to be responsible for the majority of failures. Proving that fatigue was the operative failure mode is based upon the observation of unique features visible on the fracture surface.

Fatigue is a time-dependent mechanism that can be separated into three stages. Stage I is crack initiation, Stage II is crack propagation, and Stage III is unstable fast fracture. The morphology of each stage relative to the others is unique. Figure 6-18 is a macrophotograph showing each stage.

**Fig. 6-18.    A forged 2014-T6 aluminum aircraft component that failed by fatigue.**

The fatigue crack *initiation* zone is a point (or points, giving rise to multiple origins) usually at or near the surface where the cyclic strain is greatest or where material defects or residual stresses lower the fatigue resistance of the component. The crack typically initiates at a small zone (encompassing 2-5 grains) and propagates by slip-line fracture, extending inward from the surface at roughly 45° to the stress axis (American Society for Metals, 1974c; Eyelon and Kerr, 1978). The fatigue-crack origin is therefore located at a point of local maximum stress and minimum local strength: this zone is largely determined by the component shape (including local features such as surface and metallurgical imperfections), as well as by type and magnitude of loading (American Society for Metals, 1975b). The unique role of the SEM is its ability to probe this area both morphologically and chemically to uncover any material-related factors which may have effected crack initiation.

The Stage I initiation zone appears brittle. The location of the origin is defined by interpreting features of the Stage II, or *propagation*, zone. The onset of fatigue-crack propagation is identified by a change in the orientation of the main fracture plane in each grain from one or two shear planes to many parallel plateaus separated by longitudinal ridges. The plateaus are normal to the direction of maximum tensile stress (American Society for Metals, 1974e). Stage II contains both macro- and microscopic features which directly relate to the direction of crack propagation.

Macroscopic *clamshell markings* radiate away from the origin in concentric semicircles (Fig. 6-19). These are a special form of arrest or progression marks uniquely associated with fatigue. The formation of clamshell markings has been attributed to several mechanisms, namely (1) variation in cyclic-stress levels, (2) crack-growth arrest (a period of crack retardation or rest), (3) differences in oxidation or corrosion of the fracture surface, and/or (4) slight plastic flow in the region of high stress concentration at the crack tip (American Society for Metals, 1974b and 1974e).

When the propagation zone is examined in the SEM at high magnifications, the clamshell markings can be resolved into hundreds or even thousands of *fatigue striations* (Fig. 6-19). Meyn (1966) summarized the characteristics of striations: they are always mutually parallel and at right angles to the local direction of crack propagation, they vary in striation-to-striation spacing with cyclic amplitude, they are equal in number to the number of load cycles (under stress-loading conditions), and they are generally grouped into patches within which all markings are continuous and of about the same length. In addition, fatigue striations do not cross one another, although two separate fatigue cracks may join, forming a new zone of local propagation. *Ratchet marks*, which result from multiple fatigue cracks, each producing a separate fatigue-crack zone, may be visible. As the two cracks meet, a small step is formed (Fig. 6-18).

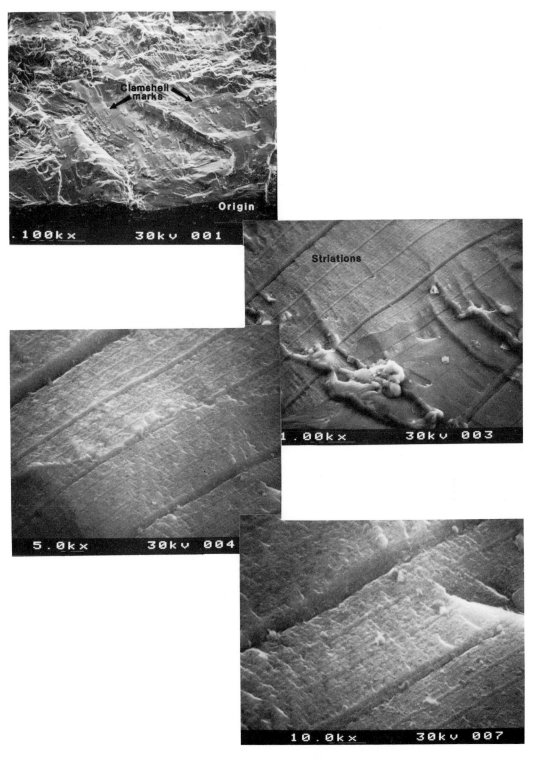

Fig. 6-19.   A series of low- to high-magnification micrographs of the specimen shown in Fig. 6-18. Note that as magnification is increased, progressively finer striations are resolved.

Because both clamshell markings and fatigue striations radiate away from the origin as a series of concentric arcs, the crack-initiation site (or sites) can be identified by drawing an imaginary radius perpendicular to their direction and centered at the origin.

Provided that the component has been subjected to uniformly applied loads, a single advance of the crack front (i.e., the distance between two adjacent striations) is a measure of the rate of crack propagation per stress cycle (Paris, 1964; McMillan and Hertzberg, 1968). This implies that one can directly relate the appearance of a fatigue fracture surface with a given stress cycle (Hertzberg and Mills, 1976; Abelkis, 1978; Madeysik and Albertin, 1978; Krasowsky and Stepanenko, 1979). However, if the loading is nonuniform, there are wide variations between a given stress-cycle series and the spacing of the striations: each stress cycle does not necessarily produce a striation. For example, overload cycles can induce microvoid coalescence, and these bands are interspersed among bands of striations (Whiteson et al, 1968). Because the area of the fracture surface occupied by dimples does not exhibit striations, there are wide variations between the pattern of striations and the applied cyclic stress (Morin and Gabriel, 1982; cf. Wiebe and Dainty, 1981). Further, under nonuniform loading, the lower load amplitudes may not be of sufficient magnitude to produce resolvable striations. In short, quantitative analysis of observed striation spacing and correlation with crack depth is difficult if not impossible to relate to nonuniform cyclic loading.

This issue is further complicated by the fact that not all fatigue fractures exhibit striations: although the presence of striations establishes that fatigue was the mode of failure, their absence does not eliminate fatigue as a possibility. Environmental conditions are one set of factors that directly affect the striations. For example, striations are well defined in aluminum alloys fatigued in air but do not form if the component is tested under vacuum (Meyn, 1968). These same conditions apply to titanium alloys (Meyn, 1971). Further, the crystallography of the alloy influences the fidelity of striations; for example, striations are often prominent in aluminum alloys but are poorly defined in ferrous alloys (Fig. 6-20). Oxidation, corrosion, or mechanical damage of the fracture surface can obliterate striations. In these circumstances, the analyst is challenged to prove that failure was primarily caused by a fatigue mechanism; proof is provided on the basis of macroscopic observations and integration of other data produced during the failure analysis.

Fractures resulting from fatigue or other modes of fracture can also exhibit structures which can be misinterpreted as fatigue striations. *Tire tracks* form when surfaces are opened, shifted, and pressed together after the crack front has passed (Phillips et al, 1965; Beachem, 1967; Koterazawa et al, 1973). Although tire tracks may be observed on fatigue fracture surfaces, they exhibit a very different morphology from stria-

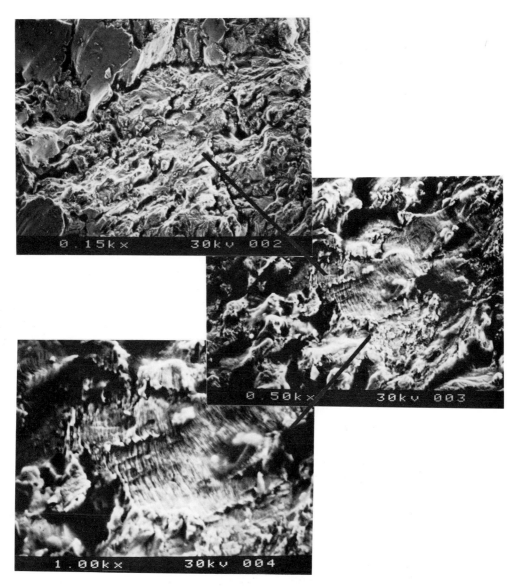

Fig. 6-20.    Poorly defined fatigue striations in a low-carbon steel specimen.

tions. *Slip lines* are steps in the fracture surface due to sheared-off layers of material and are distinguished from striations by their irregular nature (Meyn, 1966). *Wallner lines,* created by the interaction of propagating crack fronts, form a V-shaped pattern in some brittle fractures, but, unlike striations, they intersect one another (Beachem, 1968a). Linear indications or tongues on cleavage facets may be misinterpreted as fatigue striations. Lamellar structures inherent to the material's metallographic structure can also resemble fatigue striations. Ripples (or serpentine glide) are ductile, not brittle, strain lines, and as such do not fulfill the criteria required for identification of fatigue striations. Rubbing and abrasion can produce striation-like linear indications (Fig. 6-21). The microscopist must learn to distinguish these structures from striations by examining several areas across the diameter of the fracture surface. As with the other fracture modes, a series of micrographs from low to high levels of magnification must be recorded to define fatigue; in some specimens, magnifications of at least 10,000× may be necessary to resolve very fine striations (e.g., Morin and Gabriel, 1982). This microscopic data can then be inserted into the context of the entire failure analysis, and fatigue either confirmed or eliminated as the failure mode.

The fidelity of fatigue striations is enhanced by shadowing the fracture surface in the direction of crack propagation with an evaporated gold film. The gold will deposit on the leading edge of the striations, increasing their prominence and thus their resolvability. This is especially useful if magnifications exceeding roughly 7500× are necessary to resolve very fine striations. The theory and practice of shadowing are discussed in Chapter 8.

The final stage of a fatigue fracture is Stage III, *unstable fast fracture*. During crack growth, the cross-sectional area of the component is reduced. Eventually a point will be reached where the component can no longer withstand the applied load, and it separates by one of the overload mechanisms (microvoid coalescence or brittle fracture). The macroscopic and microscopic features of Stage III are therefore very different from those of the crack initiation or propagation zones.

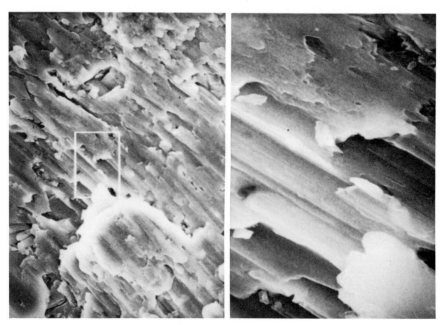

Fig. 6-21.   Striation-like lines produced by abrasion. Magnifications: 500× (left); 1000× (right).

In addition to "pure" fatigue failures, fatigue can also be combined with corrosive and thermal effects. *Corrosion-fatigue failures* arise from the combination of a corrosive environment and repeated stress. In general, the component fails sooner than if subjected to corrosion or cyclic stress alone, although exceptions have been described (American Society for Metals, 1975a). *Elevated-temperature fatigue* is encountered when the fatigue strength of a component decreases as temperature increases with concurrent cyclic loading (American Society for Metals, 1975a). This mechanism may be observed in jet engine turbine blades. In comparison, *thermal fatigue* arises from temperature cycling and mechanical restraint of the component, and may be observed in internal combustion engines and heat exchangers (American Society for Metals, 1975b). *Contact fatigue* may arise in components that roll or slide, or roll and slide, while under high contact pressure (Leonard and Meakin, 1974; American Society for Metals, 1975b). In addition to examining the fracture surface, important information can be derived by analyzing the debris from contact-fatigue failures (Seifert and Westcott, 1972 and 1973; Scott and Mills, 1974).

# REFERENCES

Abelkis, P. R. (1978) Use of microfractography in the study of fatigue crack propagation under spectrum loading. In: *Fractography in Failure Analysis*. (Strauss, B. M., and W. H. Cullen, eds.) American Society for Testing and Materials Special Publication 645, ASTM, Philadelphia, p 213.

American Society for Metals (1974a) Fractographic features revealed by light microscopy. In: *Fractography and Atlas of Fractographs*, Metals Handbook, 8th ed., vol 9, p 27. ASM, Metals Park, OH.

———— (1974b) Interpretation of light microscope fractographs. Ibid, p 36.

———— (1974c) Interpretation of scanning electron microscope fractographs. Ibid, p 64.

———— (1974d) Photography of fractured parts and fracture surfaces. Ibid, p 11.

*———— (1974e) Use of fractography for failure analysis. Ibid, p 106.

———— (1975a) Corrosion fatigue failures. In: *Failure Analysis and Prevention*, Metals Handbook, 8th ed., vol 10, p 240. ASM, Metals Park, OH.

*———— (1975b) Fatigue failures. Ibid, p 95.

*———— (1975c) General practice in failure analysis. Ibid, p 10.

———— (1975d) Hydrogen damage failures. Ibid, p 230.

———— (1975e) Liquid metal embrittlement. Ibid, p 228.

———— (1975f) Stress corrosion cracking. Ibid, p 205.

———— (1975g) Elevated temperature failures. Ibid, p 249.

Andersen, P. A., and D. J. Duquette (1980) The effects of dissolved oxygen, chloride ion, and applied potential on the SCC behavior of Type 304 stainless steel in 290C water. *Corrosion* 36(8):409.

Anderson, R. C. (1983) *Visual Examination*. American Society for Metals, Metals Park, OH.

Bandyopadhyay, N., and C. L. Briant (1983) Caustic stress corrosion cracking of NiCrMoV rotor steels — the effects of impurity segregation and variation in alloy composition. *Met. Trans.* 14A:2005.

Bandyopadhyay, N., and C. J. McMahon, Jr. (1983) The micromechanisms of tempered martensite embrittlement in 4340-type steels. *Met. Trans.* 14A:1313.

Beachem, C. D. (1963) An electron fractographic study of the influence of plastic strain conditions upon ductile rupture processes in metals. *Trans. ASM* 56(3):318.

——— (1967) Microscopic fatigue fracture surface features in 2024-T3 aluminum and the influence of crack propagation axle upon their formation. *Trans. ASM* 60:324.

*——— (1968a) Microscopic fracture processes. In: *Fracture*. (Liebowitz, H., ed.) Academic Press, New York, vol 1, p 243.

———, ed. (1968b) *Electron Fractography*. American Society for Testing and Materials STP 436, ASTM, Philadelphia.

*——— (1972) A new model for hydrogen-assisted cracking (H embrittlement). *Met. Trans.* 3:437.

*——— (1975) The effects of crack tip plastic flow directions upon microscopic dimple shapes. *Met. Trans.* 6A:377.

Beachem, C. D., and D. A. Meyn (1968) Fracture by microscopic plastic deformation processes. In: *Electron Fractography*. (Beachem, C. D., ed.) American Society for Testing and Materials STP 436, ASTM, Philadelphia, p 59.

Beachem, C. D., and R. M. N. Pelloux (1964) Electron fractography — a tool for the study of micromechanisms of fracturing processes. In: *Fracture Toughness Testing and Its Applications*. American Society for Testing and Materials STP 381, ASTM, Philadelphia, p 210.

Beachem, C. D., and G. R. Yoder (1973) Elastic plastic fracture by homogeneous microvoid coalescence tearing along alternating shear planes. *Met. Trans.* 4:1145.

*Bhattacharyya, S., et al (1979) *IITRI Fracture Handbook*. *Failure Analysis of Metallic Materials by SEM*. IIT Research Institute, Chicago.

*Boardman, B. E. (1979) Failure analysis — how to choose the right tool. *SEM, Inc.* 1:339.

Broek, D. (1974) Some contributions of electron fractography to the theory of fracture. *Int. Met. Rev.* 19:135.

Brooks, J. A., et al (1983) Effect of weld composition and microstructure on hydrogen assisted fracture of austenitic stainless steel. *Met. Trans.* 14A:75.

*Burghard, H. C., and N. S. Stoloff (1968) Cleavage phenomena and topographic features. In: *Electron Fractography*. (Beachem, C. D., ed.) American Society for Testing and Materials STP 436, ASTM, Philadelphia, p 32.

Colangelo, V. J., and F. A. Heiser (1974) *Analysis of Metallurgical Failures*. John Wiley and Sons, New York.

*Dahlberg, E. P. (1974) Techniques for cleaning service failures in preparation for scanning electron microscopy and microprobe analysis. *IITRI/SEM*, p 911.

*——— (1976) Failure analysis by examination of fracture surfaces. Analytical procedures and cleaning techniques for field failures. *ITTRI/SEM* 1:715.

*———, and R. D. Zipp (1981) Preservation and cleaning of fractures for fractography — update. *SEM, Inc.* 1:423.

DeLeiris, H., et al (1966) Techniques of de-rusting fractures of steel parts in preparation for electronic micro-fractography. *Mem. Scient. Rev. de Met.* 63:463.

Edwards, A. J. (1963) Depth measurements on fracture surfaces. Report of NRL Progress, Naval Research Laboratory, Washington, DC.

Engel, L., and H. Klingele (1981) *An Atlas of Metal Damage*. Prentice-Hall, Englewood Cliffs, NJ.

*Eyelon, D., and W. R. Kerr (1978) Fractographic and metallographic morphology of fatigue initiation sites. In: *Fractography in Failure Analysis*. (Strauss, B. M. and W. H. Cullen, eds.) American Society for Testing and Materials Special Publication 645, ASTM, Philadelphia, p 235.

Farrar, J. C. M. (1974) The role of the SEM in the failure analysis of welded structures. *IITRI/SEM*, p 859.

Gurland, J., and J. Platean (1963) The mechanism of ductile rupture of metals containing inclusions. *Trans. ASM* 56:442.

*Hertzberg, R. W., and W. J. Mills (1976) Character of fatigue fracture surface micromorphology in the ultra-low growth rate regime. In: *Fractography— Microscopic Cracking Processes*. American Society for Testing and Materials Special Publication 600, ASTM, Philadelphia, p 220.

Kayafas, I. (1980) Corrosion product removal from steel fracture surfaces for metallographic examination. *Corrosion* 36(8):443.

Knapp, B. B. (1948) Preparation and cleaning of specimens. In: *The Corrosion Handbook*. John Wiley and Sons, New York, p 1077.

Koterazawa, R., et al (1973) Fractographic study of fatigue crack propagation. *J. Eng. Mat. Tech.* 95H:202.

Krasowsky, A. J., and V. A. Stepanenko (1979) A quantitative stereoscopic fractographic study of the mechanism of fatigue crack propagation in nickel. *Int. J. Fract.* 15:203.

Kwon, H., and C. H. Kim (1983) Tempered martensite embrittlement in Fe-Mo-C and Fe-W-C steel. *Met. Trans.* 14A:1389.

Landon, P. R., et al (1983) The influence of the M (C, N) compound on the mechanical properties of type 422 stainless steel. *Met. Trans.* 14A:1395.

Lane, G. S., and J. Ellis (1971) The examination of corroded fracture surfaces in the scanning electron microscope. *Corr. Sci.* 11:661.

Leonard, L., and J. D. Meakin (1974) The application of scanning electron microscopy to studies in rolling contact fatigue. *IITRI/SEM*, p 875.

Liaw, P. K., et al (1982) Influence of corrosive environments on near-threshold fatigue crack growth in 403 stainless steel. *Met. Trans.* 13A:2177.

Madeysik, A., and L. Albertin (1978) Fractographic method of evaluation of the cyclic stress amplitude in fatigue failure analysis. In: *Fractography in Failure Analysis*. (Strauss, B. M., and J. R. Cullen, eds.) American Society for Testing and Materials Special Publication 645, ASTM, Philadelphia, p 73.

McMillan, J. C., and R. W. Hertzberg (1968) Application of electron fractography to fatigue studies. In: *Electron Fractography*. (Beachem, C. D., ed.) American Society for Testing and Materials STP 436, ASTM, Philadelphia, p 89.

Metallurgical Transactions (1983) Symposium on the role of trace elements and interfaces in creep failure. *Met. Trans.* 14A:521.

Meyn, D. A. (1966) Memorandum, report 1707. Naval Research Laboratories, Washington, DC.

*——— (1968) The nature of fatigue-crack propagation in air and vacuum for 2024 aluminum. *ASM Trans.* 61(1):52.

——— (1971) An analysis of frequency and amplitude effects on corrosion fatigue crack propagation in Ti-8Al-1Mo-1V. *Met. Trans.* 2:853.

Morin, C. R., and B. L. Gabriel (1982) The role of SEM in fatigue fracture analysis. *The Microscope* 30:139.

Nathan, C. C. (1965) Corrosion inhibitors. In: *Encyclopedia of Chemical Technology*. John Wiley and Sons, New York, vol 6, p 317.

Nielsen, N. A. (1968) Environmental effects on fracture morphology. In: *Electron Fractography*. (Beachem, C. D., ed.) American Society for Testing and Materials STP 436, ASTM, Philadelphia, p 124.

Paris, P. C. (1964) The fracture mechanics approach to fatigue. *Proc. 10th Sagamore Conf*. Syracuse University Press, Syracuse, NY, p 107.

Pelloux, R. M. (1974) Fracture mechanics and SEM failure analysis. *IITRI/SEM*, p 851.

——— (1976) Failure analysis by examination of fracture surfaces. *ITTRI/SEM*, 1:723.

*Phillips, A., et al (1965) *Electron Fractographic Handbook*. Air Force Materials Laboratory — TR-64-416, Wright-Patterson Air Force Base, OH.

Pickens, J. R., et al (1983) The effect of loading mode on the stress corrosion cracking of aluminum alloy 5083. *Met. Trans*. 14A:925.

*Pittinato, G. F., et al (1975) *SEM/TEM Fractography Handbook*. Metals and Ceramics Information Center, Battelle Columbus Laboratories, Columbus, OH.

Sadananda, K., and P. Shahinian (1983) Creep crack growth behavior of several structural alloys. *Met. Trans*. 14A:1467.

Schmerling, M., et al (1981) Auger electron spectroscopy as applied to the study of the fracture behavior of materials. *SEM, Inc*. 1:431.

Schuster, G., and C. Altstetter (1983) Fatigue of stainless steel in hydrogen. *Met. Trans*. 14A:2085.

Scott, D., and G. H. Mills (1974) Debris examination in the SEM: A prognostic approach to failure prevention. *IITRI/SEM*, p 883.

*Seifert, W. W., and V. C. Westcott (1972) A method for the study of wear particles in lubricating oil. *Wear* 21:27.

——— (1973) Investigation of iron content of lubricating oils using ferrograph and emission spectrometer. *Wear* 23:239.

*Strauss, B. M., and W. H. Cullen, eds. (1978) *Fractography in Failure Analysis*. American Society for Testing and Materials Special Publication 645, ASTM, Philadelphia.

*Tung, P. P., et al, eds. (1981) *Fracture and Failure: Analyses, Mechanisms, and Application*. American Society for Metals, Metals Park, OH.

Whiteson, B. V., et al (1968) Special fractographic techniques for failure analysis. In: *Electron Fractography*. (Beachem, C. D., ed.) American Society for Testing and Materials STP 436, ASTM, Philadelphia, p 151.

Wiebe, W., and R. V. Dainty (1981) Fractographic determination of fatigue crack growth rates in aircraft components. *Can. Aero. and Space J*. 27(2):107.

Wulpi, D. J. (1985) *Understanding How Components Fail*. American Society for Metals, Metals Park, OH.

Yuzawich, P. M., and C. W. Hughes (1978) An improved technique for removal of oxide scale from fractured surfaces of ferrous material. *Pract. Metallog*. 15:184.

*Zipp, R. D. (1979) Preservation and cleaning of fractures for fractography. *SEM, Inc*. 1:355.

*Recommended reading.

# BIBLIOGRAPHY: CERAMICS AND PLASTICS

Brockway, G. S. (1982) Resisting impact fracture of plastic parts. *Plastics Design Forum* 7(5):47.

——— (1983) Initiation and growth of cracks in plastic parts. *Plastics Design Forum* 8(1):69.

Cross, P. M. (1974) The tensile failure of thermoplastics. An investigation using the scanning electron microscope. *IITRI/SEM*, p 995.

Mecholsky, J. J., et al (1978) Fractographic analysis of ceramics. In: *Fractography in Failure Analysis*. (Strauss, B. M., and W. H. Cullen, eds.) American Society for Testing and Materials Special Publication 645, ASTM, Philadelphia, p 363.

Sklad, P. S., and J. Bentley (1981) Analytical electron microscopy of TiB-Ni ceramics. *SEM, Inc.* 1:177.

Williams, J. G. (1981) Fracture mechanics of nonmetallic materials. *Phil. Trans.* (Roy. Soc. London), Series A, 299:59.

# =7=

# Replicas

*Replicas* are exact reproductions of surfaces which may be microscopically examined when the original specimen cannot be studied directly. A very common application is the study of specimens that are too large to fit in the SEM specimen chamber. This situation is frequently encountered during failure analyses involving litigation, for example, when sectioning of the failed item is prohibited. In other situations it may be difficult to transport a failed structure from the field to the laboratory, but replicas may be prepared in the field and later examined in the laboratory. The protection of fracture surfaces from corrosion and oxidation by sealing with replicas was discussed in Chapter 6. Replicas are also used in fractography to clean the surface of interest, and extraction replicas preserve the distribution of the removed materials.

Replicas may also be used to record the progression of surface defects under experimental conditions. For example, in studies of gear wear, a minute surface imperfection may evolve into a pit or spalled area because of rolling-contact fatigue failure. To remove a gear after a prescribed cycling, examine it by SEM, and then resume cycling is inefficient. Rather, a series of replicas stripped after a predetermined number of cycles (e.g., every 1000 cycles) can be used to trace the final failure back to its initiation stage (Anderson, 1974). Such chronological records are useful in any type of wear study.

Discussed below are the materials and methods commonly used in the replication of materials for SEM. Since replicating in materials science is a hybrid methodology which arose from TEM technology and dentistry, the reader should not be surprised to find several dental journals cited in the references. Relatively few publications uniquely devoted to replication in materials science have appeared in the technical literature, although its usefulness is well established (Warke et al, 1968; Pelloux, 1970; American Society for Metals, 1974; Phillips et al, 1976). Good reviews of general replicating techniques are available in Barnes (1972 and 1978), Pfefferkorn and Boyde (1974), Pameijer (1975 and 1978), and Kusy and Leinfelder (1977).

## MATERIALS AND METHODS

*Single-stage replicas* prepared from cellulose acetate or dental impression media are most commonly used in SEM. Because the impression is completed in one step, single-stage replicas are *negative copies*. When a negative copy of the specimen is examined, the image produced will be an exact but opposite copy of the specimen; for example, pits in the specimen will appear as mounds on the replica (Fig. 7-1). Image inversion, a signal-processing device discussed in Chapter 2, will electronically re-establish the original perspective and prevent misinterpretation of surface relief. Alternatively, two-stage replicas are prepared by replicating the negative impression, which produces a *positive replica*.

Materials used as replicating media must accurately reproduce the original surface. This implies that the medium must have good wetting properties (or low surface tension), which will produce a low contact angle between the medium and the surface (Pameijer and Stallard, 1973; Pameijer, 1975 and 1978). Other characteristics that a good replicating medium possesses are: rapid polymerization without shrinkage, easy detachment from the surface without distortion, and resistance to vacuum, irradiation, and heat. Many dental impression media and cellulose acetate fulfill these criteria. Other replicating techniques employ evapo-

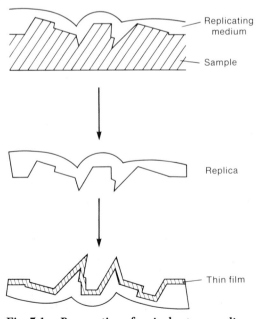

Fig. 7-1.   Preparation of a single-stage replica.

rated coatings; these methods are more commonly used in transmission electron microscopy and will not be discussed here. Interested readers are referred to Bradley (1967), Phillips et al (1976), and Willison and Rowe (1980).

## Cellulose Acetate

*Cellulose acetate* replicating tape is available in roll form from electron microscope suppliers. The tape is 1″ wide and of varying thickness; tape 0.127 mm thick is used for reproducing rough surfaces, while 0.022-mm tape is for finer surfaces. The tape width obviously limits the area which can be accurately replicated; either successive, overlapping (montage-like) replicas may be prepared, or *acetylcellulose sheets* (10 cm × 12 cm with thickness of 0.08 mm or 0.034 mm) may be used to replicate large areas.

The replica is prepared using the "pressure" technique as follows: the specimen surface is cleaned with compressed gas and a solvent, if necessary, and dried. It is then visually examined and sites are chosen for replication. The surface is photographed or sketched, and the replica site is indicated. A section of tape, roughly half an inch longer than necessary, is moistened on one side with a small volume of acetone. After 7-10 sec, the tape is positioned over the surface and firmly held in place with moderate finger pressure for about 2 min. The pressure is released and the tape dried in place for 10-15 min. The tape will solidify into a plastic-like sheet which can be lifted from the original surface. If the replica retains surface debris, additional replicas should be taken until a clean one is obtained. Finally, the clean replica is carefully stored in a cool area.

The analyst must follow several rules for this method to be successful. To avoid reproducing one's fingerprint on the opposite side of the replica, only one side of the tape should be dissolved. The tape should not be too soft, which will give rise to bubbles, and should not be too hard, which will interfere with good contact and result in smooth, featureless areas. A small-diameter pipet is used to control the amount of acetone applied to the tape. When positioning and holding the tape over the specimen surface, do not shift or readjust its position. Any movement will smear the replica. Apply a moderate amount of pressure to force the cellulose acetate into fine surface features. Excessive pressure thins the replica, which will consequently tear when it is removed, whereas insufficient pressure results in poor contact. The combination of heat and pressure produced when using the thumb or finger to hold the replica produces a better copy than pressure alone. Because two minutes is a relatively long period of time to remain motionless, the analyst should be in a comfortable position. If rough areas of the fracture protrude through the replica, a second replica can be applied over the first to make the resulting composite replica thicker and stronger.

While the replica is solidifying, indicate its position and orientation on a sketch or photograph. This is essential to relate surface features to the macroscopic model; for example, to determine the directionality of a fracture, one must know the orientation of the replica relative to the original fracture surface. Simply notch a corner of the cellulose acetate tape at the noncontact zone, and use this as a reminder. Preliminary visual examination of a fracture surface usually indicates the origin of the fracture; using a waterproof ink, indicate the directionality with arrows on the replica. Though these practices are always important, they become critical when the microscopist does not have access to the original surface and cannot directly compare the replica with the original surface. Without detailed sketches and labels, replicas are difficult to interpret.

After curing for 10-15 min, the hardened replica is stripped from the surface. The replica must be completely dry to avoid tearing, smearing, or scratching artifacts. At a shallow angle, pull the replica up and away from the surface; do not use any sideways motion or recontact the original. The pulling motion should be gentle and continuous. If the replica is still soft, do not reposition it over the surface; rather, wait for it to dry, discard it, and start over.

The stripped replica is then examined with a low-power binocular microscope to determine whether any artifacts have arisen during preparation. *Artifacts* are imperfections due to poor technique which are not part of the original surface (Dahlberg and Beachem, 1965; Phillips et al, 1976). They are most easily identified in a side-by-side comparison of the original surface with the impression. If the original specimen is not available, the microscopist must learn to discriminate between real structures and artifacts; the differences can be subtle. Novices should practice replicating a specimen surface that can be examined directly in the SEM, and compare the original with the replica. With practice, accuracy will increase.

Bubbles, the most common artifact seen in replicas, arise when the tape has been allowed to get too soft. Using excessive acetone and/or waiting too long before positioning the replica on the surface produce a too-soft tape. The resulting bubbles are clearly visible in the binocular microscope and interfere with SEM imaging of the fine structure. Use of excessive acetone will also dissolve both sides of the tape, and during impression a two-sided replica will form: on one side is the surface of interest, and on the other side is a thumbprint.

An excessively soft tape and/or excessive pressure during application can cause high points of the original surface to poke through the replica, whereas insufficient pressure produces featureless regions resulting from lack of contact. Any movement during the initial curing phase will smear the replica. Tearing or local strain results if the replica is stripped before complete drying; always wait at least 10 min before stripping, and increase this time on hot or humid days. This is also important when cleaning a fracture surface with extraction replicas; recall from Chapter 6 that cellulose acetate adheres to rough surfaces and interferes with SEM imaging.

Artifactual "striations" may be observed if the replica has been improperly stripped, scraping the replica on the original surface. Using one continuous motion to remove the replica eliminates this problem.

When a satisfactory replica has been obtained, it is then trimmed, mounted on a stub, and coated with a metal thin film for SEM examination. Scissor cutting can produce "striations" or local strain; it is better to use one perpendicular razor cut to trim the replica (do not "sawcut" the replica). The trimmed replica is mounted on a stub using a small amount of conductive paint. The paint should contact only the underside of the replica, although the best connection to ground is made by painting a continuous, thin stripe of paint from the stub to the replica surface. Obviously the paint should not be applied to the replicated surface, but only to an area not containing the surface of interest. The electron density of the specimen is then increased by coating with a thin layer (~200 Å) of metal (usually gold). The thin film is applied using either the rotary evaporation or sputter coating techniques, which are discussed in Chapter 8. Overheating during coating can melt, wrinkle, or crack the replica. Artifacts due to overheating may also arise during SEM examination, and are caused by a large spot size or excessive accelerating voltage. Always examine cellulose acetate replicas with a small-to-moderate beam diameter and below 20 keV.

## Dental Impression Media

*Impression media* are extremely important in restoration dentistry, and very sensitive elastic impression materials have been developed. Materials scientists have adapted these materials and methods to replicate surfaces which are not reproducible with the cellulose acetate method. For example, if a very large surface is to be reproduced, it is tedious to accurately reconstruct that surface by many overlapping cellulose acetate replicas. Rather, a single, large replica may be prepared using a dental impression medium. Another application involves the reproduction of extremely rough or hard-to-reach surfaces. Accurately reproducing the contour of a bolt hole, for example, would be next to impossible with tape. However, it is easy to fill such a cavity with the impression monomer, which will polymerize but remain sufficiently elastic to be easily removed. Dental impression media are also useful if the original specimen is sensitive to acetone; clearly, one does not want to destroy the original specimen during replication.

A large number of elastic impression media are available; the scanning electron microscopist is most interested in those which accurately reproduce surfaces with minimal artifacts. Craig et al (1975) thoroughly discuss the media and their properties, and Pameijer (1979) evaluates several media using the SEM. Since only a few media are actually required by the materials scientist, these replicating methods are greatly simplified.

A given medium is usually supplied in separate tubes of a base and a catalyst, and the working medium is prepared by mixing the two components in the recommended proportions. Thorough mixing of the recommended volumes is critical to success, because otherwise the medium will not polymerize. The mixing must be done rapidly because the catalyst quickly initiates polymerization, and viscosity increases. However, mixing that is too vigorous will introduce air bubbles into the monomer. A simple but effective mixing method uses a clean glass sheet and a small spatula. After identifying and cleaning the surface to be replicated, prepare the medium by placing the specified proportion (usually equal parts) of base and catalyst next to each other on the glass sheet. Mix the two components by smoothing one into the other using the flat end of the spatula. Because the base and catalyst are of different colors, inadequate mixing is visible as streaking. Do not stir the medium, which will introduce air bubbles, and do not extend the mixing time beyond that necessary to homogenize the medium because it will begin to polymerize and increase in viscosity. Using the pressureless technique, immediately apply the medium to the original surface and smooth it with the spatula. Gentle pressure ensures good contact between the monomer and original. Place a liberal amount of the monomer on the surface; an excess amount is easier to handle than too little. The medium will cure into an elastic polymer in 10-15 min. It is then lifted from the surface, labeled, and stored away from heat. As with cellulose acetate replicas, careful sketches or photographs of the original surface must be prepared, and the orientation of the replica indicated. These replicas are bulkier than those prepared with cellulose acetate and may be difficult to secure on a stub. A method developed by Lametschwandtner et al (1980) for securing biological vascular casts (analogous to replicas) works very well. They glue two or more fine copper wires to a stub with silver paint and then press the replica onto the wires. This simultaneously secures the specimen and establishes a good contact to ground. For general purposes, the author uses Accoe* silicone impression medium, which is easily prepared, has moderate viscosity, does not introduce significant distortion, and is readily coated with a metal thin film (Fig. 7-2). Other elastic impression materials such as Kerr[†] Permelastic and Reflect media are useful, but require a preliminary carbon coating before the metal thin film is applied; used alone, metallic films will only form a continuous film with very heavy deposits.

---

*Available from Coe Laboratories, Inc., Chicago, and other suppliers.
[†]Available from Frank Dental Supply, Elk Grove Village, IL, and other suppliers.

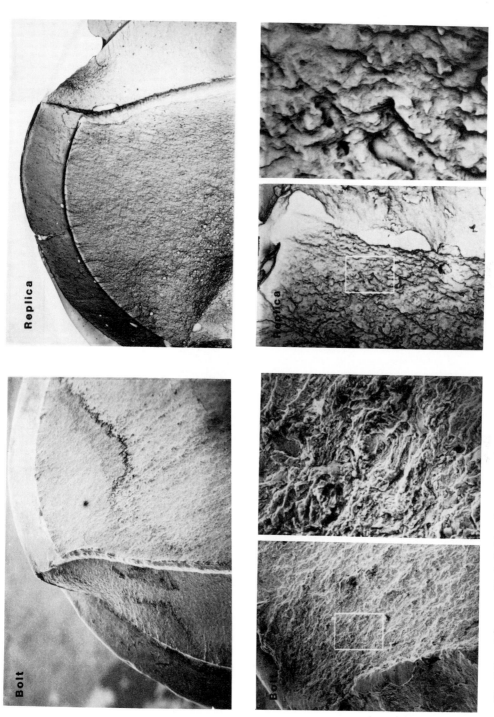

Fig. 7-2. A fractured bolt (left) and an inverted image of a silicone replica (right). Note that the bolt and replica are mirror images. Upper micrographs at 20×; dual-magnification micrographs at 100× and 500×. All micrographs shown here at 75%.

When filling a cavity such as a bolt hole, it is sometimes easier to introduce the mixed monomer with a disposable syringe. Simply fill the syringe with monomer, introduce its tip into the cavity, and fill until overflowing. The advantage of using a syringe is that the number of bubbles in the polymerized cast is reduced, but the disadvantage is that unless one works quickly, the monomer viscosity increases and a poor impression results.

The artifacts associated with dental impression replicas usually arise from improper mixing of the final medium. The proportion of base to catalyst as specified by the manufacturer must be used to ensure complete polymerization. Insufficient catalyst slows (or eliminates) the rate of polymerization, while excess catalyst accelerates polymerization and increases viscosity, producing a medium that is too solid to accurately reproduce fine surface features.

Bubbles are also the most common artifact seen in replicas cast from dental media. Some are sufficiently large to be visible to the naked eye, whereas others require increased magnification. In the SEM, bubbles are manifested as very smooth spherical or ovoid regions, and burst bubbles are visible as cavities. Either form interferes with the duplication of surfaces. Bubbles arise from too vigorous stirring of the medium, from insufficient smoothing of the monomer over the surface, and from attempting to replicate a wet surface. If the mixed monomer does not rapidly polymerize, the medium may be outgassed with a mild vacuum, but be aware that viscosity will increase during this interval.

## Comparison of Casting Media

The choice of a dental impression medium or cellulose acetate is based upon several factors. When very fine structures, resolvable only beyond 1000×, are to be imaged, cellulose acetate is the better choice. It will provide better resolution at higher magnification than dental media, which are limited to ~500× magnification. The resolution limit is largely limited by the surface tension and flow properties of the medium; cellulose acetate has low surface tension and better flow properties than do dental media. On the other hand, the dimensions of an area to be reproduced may be hampered by the limited size of cellulose acetate tape; dental media may be preferred for replicating large areas. The latter are also preferred for reproducing inaccessible surfaces.

Both types of replicas are coated with a thin conductive film in preparation for SEM examination. During SEM evaluation, each medium is identically treated. The coated replicas are examined with a moderate accelerating voltage (10-20 keV) and a relatively small spot size. Higher voltages should be avoided, because virtually all data are being emitted from the outermost, gold-coated layer of the specimen. Higher voltages will increase the depth of penetration, and because the replica itself is nonconductive, charging artifacts may arise.

Fig. 7-3.    A cellulose acetate replica recorded in the negative (a) and positive (b) imaging modes. The area shown is similar to that in Fig. 7-2. Magnifications: 100× (left); 500× (right).

A disadvantage in examining replicas is that differential atomic number contrast (as observed in metals) is absent. The analyst can extend the limits of the gray scale using gamma or black-level subtraction as discussed in Chapter 2. The inverted image signal-processing mode is also helpful when examining single-stage replicas; this mode produces a negative of a negative replica, the net effect being the display of a positive image. Although this is not as accurate as imaging of the original surface, it effectively restores the orientation of displayed features. If the SEM does not have the inverted imaging mode, counterpositives (negative prints) may be prepared using conventional darkroom techniques. Reference to Fig. 7-3 may assist the confused reader.

A completely different approach is the preparation of a *positive replica*. Two-stage replicas are prepared by impressing a negative replica and using this as a template for a second, or positive, replica. The obvious advantage of positive replicas is that they reproduce the orientation and perspective of the original specimen. The accuracy lost between the two replicas under ideal conditions is roughly 2%: the positive replica is only as good as, not better than, the negative replica.

Pameijer (1975) described a relatively simple method for preparing two-stage replicas. First, the surface is replicated with cellulose acetate. A layer of a commercial parting agent (e.g., Victawet) is evaporated over the replica surface, followed by a heavy gold deposition. The replica is stabilized by first electroplating a thin layer of copper over the gold, followed by a layer of self-curing resin. The gold-copper-resin laminate is separated from the cellulose acetate and is the positive replica. Other, lengthier procedures are usually associated with transmission electron microscopy; refer to the earlier citations for these methods.

## REFERENCES

American Society for Metals (1974) *Fractography and Atlas of Fractographs,* Metals Handbook, 8th ed., vol 9. American Society for Metals, Metals Park, OH.

Anderson, S. (1974) Plastic replicas for optical and scanning electron microscopy. *Wear* 29:271.

Barnes, I. E. (1972) Replica models for the scanning electron microscope. *Brit. J. Dent.* 1:333.

*—— (1978) Replication techniques for the SEM. I. History, materials, and techniques. *J. Dent.* 6:327.

Bradley, D. E. (1967) Replica and shadowing techniques. In: *Techniques for Electron Microscopy,* 2d ed. (Kay, D. H., ed.) Blackwell, Oxford, p 96.

Craig, R. G., et al (1975) *Dental Materials, Properties, and Manipulations.* C. V. Mosby Co., St. Louis.

Dahlberg, E. P., and C. D. Beachem (1976) Some artifacts possible with the two-stage plastic-carbon replication technique. *NRL Rpt.* 1457.

Kusy, R. P., and K. F. Leinfelder (1977) In situ replication techniques: I. Preliminary screening and the negative replication technique. *J. Dent. Res.* 56:925.

Lametschwandtner, A., et al (1980) On the prevention of specimen charging in the SEM of vascular corrosion casts by attaching conductive bridges. *Mikroskopie* 36:270.

Pameijer, C. H. (1975) Replica techniques. In: *Principles and Techniques of Scanning Electron Microscopy,* vol 4. (Hayat, M. A., ed.) Van Nostrand Reinhold, New York, p 45.

*—— (1978) Replica techniques for scanning electron microscopy—a review. *SEM, Inc.* 2:831.

—— (1979) Replication techniques with new dental impression materials in combination with different negative impression materials. *SEM, Inc.* 2:571.

Pameijer, C. H., and R. E. Stallard (1973) Three replica techniques for biological specimen preparation. *IITRI/SEM,* p 357.

Pelloux, R. M. (1970) Replicating techniques for electron fractography. In: *Applications of Modern Metallographic Techniques.* American Society for Testing and Materials STP 480, p 127.

Pfefferkorn, G. E., and A. Boyde (1974) Review of replica techniques for scanning electron microscopy. *IITRI/SEM,* p 75.

*Phillips, A., et al (1976) *Electron Fractography Handbook* AFML TDR 64-416. Battelle Columbus Laboratories, Columbus, OH.

Warke, W. R., et al (1968) Techniques for electron microscopic fractography. In: *Electron Fractography.* American Society for Testing and Materials STP 436, ASTM, Philadelphia, p 212.

Willison, J. H. M., and A. J. Rowe (1980) Replica, shadowing, and freeze etching techniques. In: *Practical Methods in Electron Microscopy,* vol 8. (Glauert, A. M., ed.) American Elsevier, New York.

---

*Recommended reading.

# =8=

# Thin Films

Examination of uncoated, nonconductive specimens in the SEM is difficult because the specimens behave like insulators by absorbing electrons, and accumulate a net negative charge. Consequently, the specimen deflects the electron beam (like charges repel like charges) and image quality is degraded. Although one can reduce these charging artifacts by lowering the accelerating voltage and spot size, magnification and resolution are limited (refer to Chapter 1 and Shaffner and Hearle, 1976). The image quality may be slightly improved by treating the specimen surface with organic antistatic agents, but magnification is still limited (Sikorsky et al, 1986; Pfefferkorn, 1973; Pease and Bailey, 1975; Echlin, 1981). The best method for increasing the secondary electron yield and improving image quality over the entire magnification range of the SEM is to deposit a conductive thin film over the specimen surface. These thin films (~200 Å thick) greatly improve the point-to-point resolution of a given specimen without suppressing fine surface features.

The two methods commonly used in SEM for thin-film preparation are thermal evaporation and sputter coating. Both of these methods achieve the same ends, but through very different mechanisms. Metallic thin films (e.g., gold) may be prepared using either technique; these films serve to increase the secondary electron yield and improve image quality. Carbon thin films are prepared by thermal evaporation, and improve specimen conductivity without significantly enhancing the secondary electron yield. Because the atomic weight of carbon is low, carbon films will not excessively suppress the characteristic X-rays released from the specimen; carbon films are used when an X-ray analysis of a nonconductive specimen is required. Each of these techniques and their applications are discussed below.

# THERMAL EVAPORATION

Evaporated thin films are prepared under high vacuum by passing a current through a metal wire. As the temperature of the wire increases, the metal will vaporize (at its unique *vaporization temperature*), and the released atoms follow a line-of-sight trajectory until they are intercepted by a surface within the vacuum chamber. As more metal is vaporized, a continuous thin film will grow. This method, known as *resistance heating* or *thermal evaporation*, was introduced in transmission electron microscopy by Williams and Wyckoff (1946) to increase the electron contrast of transparent specimens. Thermal evaporation has been adopted by scanning electron microscopists for several purposes. First, the imaging of nonconductive specimens is improved when the evaporated thin film is continuous. Evaporation from a point source onto a rotating and tilting specimen will produce a continuous thin film; this modification is referred to as *rotary evaporation*. The line-of-sight nature of thermal evaporation is used to advantage in *shadowing* (synonym: shadow casting; Smith and Kistler, 1977; Smith and Ivanov, 1980; Willison and Rowe, 1980). Shadowing produces an oblique deposition of the evaporant over the stationary sample, thereby highlighting surface features. For example, the fidelity of minutely spaced fatigue striations is enhanced by shadowing the fracture surface in the direction of crack propagation. Resistance heating, similar to rotary evaporation, is also used for the preparation of *carbon thin films*, which are used to enhance conductivity without sacrificing resolution in X-ray analyses. Carbon and metal thin films are normally deposited as continuous films on nonconductive specimens using rotary evaporation, whereas shadowing, for our applications, produces a discontinuous film and is preferred for conductive specimens.

## Vacuum Bell Jar

The apparatus used for thermal evaporation is the vacuum bell jar (Fig. 8-1). It consists of a large glass jar evacuated by a rotary pump and a diffusion pump connected in series; the arrangement of the vacuum system is identical to that of scanning electron microscopes having diffusion pumps. Unlike the SEM vacuum system, most bell jar vacuum systems are operated manually; the analyst must follow a given sequence of steps to obtain high vacuum. The bell jar vacuum system is controlled by valves between the (1) rotary pump (RP) and diffusion pump (DP), (2) RP and bell jar, (3) DP and bell jar, and (4) a bleed-in valve between the bell jar and atmosphere. The RP is used to rough out the bell jar from atmospheric pressure to $10^{-2}$-$10^{-1}$ torr and also as a backing pump for the DP. The DP achieves high vacuum ($10^{-5}$-$10^{-4}$ torr) from low vacuum. The relative vacuum conditions are monitored with a thermocouple gauge at backing pressures and a discharge gauge under high-

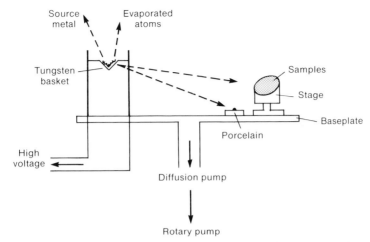

**Fig. 8-1. Cross section of the vacuum bell jar arranged for thermal evaporation.**

vacuum conditions. The general operation of a bell jar vacuum system is described under "Method of Evaporation."

The baseplate of the bell jar has several insulated ports for passage of paired high voltage cables. The circuit is completed via a conductive substrate which holds the evaporant. The holder is usually a tungsten wire shaped into a basket; because tungsten is a refractory metal, it will not evaporate, although it will alloy with some metals during evaporation. Tungsten baskets are commercially available, or may be handmade using pure wire 0.5 to 1.0 mil in diameter and a wood screw. First, heat the tungsten wire in a Bunsen burner to render it more malleable (Bradley, 1967). Wrap the wire around the threads of the screw, leaving an excess length (~2″) on both sides of the basket, and remove the screw. Each end of the wire is inserted into a matched pair of electrodes and secured. With repeated use, baskets become brittle and will eventually break, and must be replaced.

Other types of substrates are available for various applications. Tungsten boats are used to hold molybdenum apertures during cleaning, and carbon is evaporated using spring-held carbon electrodes in place of the tungsten basket.

For rotary evaporation, the specimens are placed on a rotating/tilting stage positioned at least 10 cm from the source. Because thermal evaporation follows a line of sight, the stage motion ensures that the specimen will be coated with a continuous, equally thick deposition of metal (Fig. 8-2a). If the specimen is held stationary and at an angle relative to the source, the evaporated metal will accumulate in front and on top of any prominent surface feature, while the area behind the feature remains free from evaporant (Fig. 8-2b). This is the *shadowing technique* and is used to increase the prominence of very fine surface features. Because these films are discontinuous, they are deposited on naturally

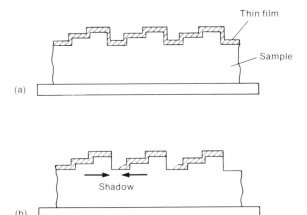

**Fig. 8-2.** Deposition of a continuous film (a) and a shadowed film (b) on a fatigue fracture (side view).

conductive specimens or on nonconductive specimens that have been precoated with a thin carbon film. If a rotating/tilting stage is unavailable, a continuous conductive layer may be obtained by coating the specimen twice; the specimen is reoriented by 180° relative to the source between each evaporation. This *portrait technique* does not produce coatings as continuous as those produced during rotary evaporation and should be used only if a planetary-motion stage is not available.

In either situation, the specimen must be a minimum of 10 cm from the source to prevent heat damage. The source will exceed temperatures of 1000 °C during evaporation, and this thermal energy is carried by atoms ejected from the source. The atoms will cool during travel, but the need for adequate separation of the source and target is clear.

The source metals commonly used for evaporation are gold, palladium, platinum, alloys of these metals, or carbon-metal mixtures. The parameters for choosing a metal will be discussed; for simplicity at this point, assume that a 5-7 cm length of the metal wire is compacted and placed in the tungsten basket. The specimens are positioned in the bell jar and evacuated to a minimum of $10^{-4}$ torr. The high voltage is turned on; then a current is passed through the basket and gradually increased until the vaporization temperature of the source metal is reached. Atoms will be ejected from the source and recondense as a thin film on all surfaces within a line of sight from the source. After the desired film thickness is achieved, the bell jar is brought to atmospheric pressure and the specimens are removed for SEM examination. The specimens must be protected from heat and humidity if they cannot be promptly studied. Presented below are the finer points of thin-film theory and a detailed description of film preparation.

## Theory of Evaporation

The following discussion of thin-film theory is a compilation of data from several sources (Shiflett, 1968; Holland, 1970; Neugebauer, 1970;

Nagatani and Saito, 1974). It will quickly become apparent that an understanding of this theory is required for its successful application in the laboratory. A simplified version of the theory of thin-film growth and the characteristics of such films are presented below.

Assume that the source metal and specimens are in position and that the bell jar is under high vacuum. As a current is passed through the tungsten basket, temperature will increase until at the vaporization temperature individual atoms are ejected from the source. Because evaporation is conducted under high-vacuum conditions, the ejected atoms follow a line-of-sight trajectory away from the source. The atoms are intercepted by all exposed surfaces within the bell jar, which includes the sample surface. Because the atoms possess significant thermal energy, they can travel considerable distances away from the source. Various mechanisms have been proposed for the interception of these atoms by a surface; in all probability, a combination of mechanisms operates simultaneously. The extreme temperature difference between the source atoms and the target causes a rapid heat exchange, and once the excess energy is dissipated, the atoms cannot travel. Henderson and Griffiths (1972) note that high-melting-point metals (such as those used here) have strong internal binding characteristics and quickly return to the ground state by entrapment at surface binding sites. In thin-film terminology, these entrapped atoms are referred to as nuclei and function as binding sites for atoms subsequently ejected from the source (Hodgkin and Murr, 1974). Nuclei are established across the specimen surface and grow as more atoms are entrapped. As more metal is evaporated, the nuclei grow into clusters and coalesce with adjacent nuclei, thus producing a thin film.

The thickness and roughness of the thin film are important because they control the limit of resolution. A film that is too thin may actually consist of discontinuous clusters of metal, which do not conduct electrons as efficiently as a continuous film (Morris and Coutts, 1977), and an overly coarse-grained or thick film obscures surface features. A very thin but continuous film which exhibits minimal graininess is desired; achieving this result is primarily a function of the characteristics of the source metal and secondarily the conditions of evaporation.

As stated earlier, the metals commonly used for thermal evaporation are gold, platinum, palladium, alloys of these metals, or carbon-metal mixtures (e.g., gold-carbon or platinum-carbon). All exhibit good thermal conductivity and high secondary electron emission, and all are easily evaporated by resistance heating. Thin films of these metals may be differentiated on the basis of minimum film thickness required for continuity and relative grain size. As a rule, the higher the melting point of the metal, the finer the thin film. As shown in Table 8-1, gold has the lowest melting point relative to other common evaporant metals, whereas platinum has the highest melting point. Therefore, platinum can form a much finer film than gold, because platinum will form a larger

Table 8-1.   Materials Used for Thermal Evaporation

| Element | Mean atomic number | Melting point, °C | Vaporization temperature, °C | Granulation |
|---|---|---|---|---|
| C . . . . . . . . . . . . . . . . . | 6 | 3800 | 2681 | Amorphous |
| Au . . . . . . . . . . . . . . . . | 79 | 1063 | 1465 | Coarse |
| Pd . . . . . . . . . . . . . . . | 46 | 1550 | 1566 | Moderate |
| Pt. . . . . . . . . . . . . . . . . | 78 | 1755 | 2090 | Fine |
| Au:Pd (60:40) . . . . . . . . | 66 | — | — | Coarse |
| Pt:Pd (80:20). . . . . . . . . | 72 | — | — | Fine |
| Pt:C (60:40) . . . . . . . . . | 50 | — | — | Very fine |

number of smaller crystallites per unit area. Because the number of small crystallites is increased, the film roughness is considerably lower. Alloyed metals (Molcik, 1967) or carbon-metal mixtures also decrease film granularity and minimize thickness by preventing migration of metal atoms through the crystal lattice (i.e., the amorphous character of the film is increased). This technique was introduced by Bradley (1958a, 1958b, and 1959) using carbon and platinum, and was modified by Harris (1975), who devised an apparatus (adaptable to any bell jar system) which controls the ratio of carbon to metal evaporated. For routine purposes in SEM, gold is adequate, but in applications requiring a resolution better than ~100 Å, an alloy or platinum is recommended.

Extremely fine films may be prepared from very-high-melting-point (refractory) metals. Refractory metals cannot be vaporized by thermal evaporation, primarily because a substitute for the tungsten basket is unavailable. Thin films of these metals are usually prepared by evaporation from an electron gun (Hagler et al, 1977; Slayter, 1978). Tungsten (Slayter, 1976), tantalum-tungsten (Abermann and Salpeter, 1974), and platinum-tungsten (Slayter, 1980) are examples of refractory metals and alloys that have been successfully evaporated. These metals produce very fine films and are used in extremely high resolution work. Electron gun evaporation is currently restricted to research situations; the interested reader should consult the references cited above for additional information.

The secondary parameters which influence thin-film fineness are the conditions of evaporation. Film thickness and graininess are reduced by evaporating under high vacuum (preferably in the $10^{-7}$ to $10^{-6}$ torr range), limiting the total deposition, lowering the rate of deposition, decreasing the target (sample) temperature, increasing the angle between the source and sample, and minimizing the contamination rate (Glang, 1970; Echlin, 1981). With regard to the relative vacuum, as a rule finer films are produced at higher vacuums. Although an adequate film may be produced at $10^{-4}$ torr, modern bell jar systems are capable of quickly achieving vacuums in the $10^{-7}$ to $10^{-6}$ torr range, and this higher region should be used to minimize film graininess. Limiting the amount of material to be evaporated obviously limits the film thickness, simply because a given amount of vaporized metal can only be spread so thin. A slower rate of deposition will also decrease film granularity (cf. Echlin, 1978): if the source metal is instantaneously evaporated, clumps of atoms tend to be released from the source, whereas at a slower rate discrete atoms or small clusters will be ejected. Baumeister and Hahn (1978) note that applying the current in pulses rather than continuously reduces film granularity. A similar technique is to deflect the evaporated atoms from a shield before interception by the specimen surface (Johansen, 1974).

Most thin-film researchers agree that lowering the temperature of the target enhances the fineness of the thin film (Echlin and Kaye, 1979; Slayter, 1980), although Filshie and Beaton (1980) disagree. Slayter (1976) noted that the excess thermal energy carried by the evaporated atoms is rapidly dissipated by the cooler specimen, reducing the surface mobility of the evaporated atoms. The net effect of cooling is to produce smaller and more uniformly sized nuclei per unit area. Further, heat-sensitive specimens may be protected by cooling (Zingsheim et al, 1970; Rowsowski and Glider, 1977; Slayter, 1980).

Contamination is a problem both during and after evaporation, but the degree of contamination is controlled by the analyst. Artifacts arising during coating and due to contamination include the entrapment of gaseous and hydrocarbon vapors; after coating, gas adsorption and settled dust degrade the film (Blaschke, 1980; Echlin, 1981). Contaminants arise from outgassing of the specimen, gaskets, and sealants during evacuation and from backflow of the diffusion pump oil. Bell jars equipped with cold traps will remove most vaporous contaminants (Echlin, 1975b), but only if the bell jar is regularly cleaned. During each evaporation cycle, a thin film is deposited on all exposed surfaces, and when the bell jar is brought to atmosphere a layer of gas and water vapor is adsorbed. As the system is recycled, gas will evolve from that surface, and another metal coating tends to produce a metal-gas-metal sandwich. The entrapped vapor will outgas during subsequent cycles, and the maximum vacuum attainable deteriorates. This problem is reduced by thoroughly cleaning all internal surfaces of the bell jar with a soft cloth and pure acetone, followed by careful drying. If the bell jar gaskets are greased, apply only a very small amount of grease to avoid outgassing and ensure vacuum integrity. The bell jar should also be pumped down to a minimum of $10^{-2}$ torr before shutting down the system.

## Method of Evaporation

For simplicity, described below is the thermal evaporation of gold onto a nonconductive specimen in order to enhance the imaging of that specimen. The other nonrefractory metals and alloys discussed earlier may be substituted for gold; they are evaporated simply by increasing the current passed through the tungsten basket until the vaporization temperature of that metal is reached. Because many bell jar systems are manual, the details of proper operation are described here; the novice should also consult the manufacturer's operating manual.

The first stage of thermal evaporation is to turn on the vacuum system. Check that all the vacuum valves are in the closed position. Turn on the diffusion pump cooling water, then the main power switch and the rotary pump. Wait roughly 30 sec (until the rotary pump quiets), open the valve between the rotary and diffusion pumps such that the diffusion pump is backed by the rotary pump, and turn on the diffusion pump. A period of 20-30 min is required for the diffusion pump oil to heat, and during this time the specimens may be cleaned and mounted on stubs. If a conductive paint is used as the adhesive, make sure that the paint has completely dried before proceeding to the next step.

After the diffusion pump oil has heated, air is admitted to the bell jar by opening the bleed-in valve. Samples to be coated with a continuous thin film are mounted on the rotating/tilting stage, and the stage is placed on the baseplate 10-15 cm away from the source. Do not position the specimen stage directly beneath the source; molten metal may drip from the source onto the specimen. For shadowing, the specimens are positioned on a slanted (10-30°) stage that is positioned along a line of sight from the source. Also position a clean white porcelain plate near the specimens, and place a small drop of vacuum oil on the plate. This is a crude but effective means to estimate film thickness; the area beneath the drop remains white while the surrounding area becomes darker as film thickness increases. Because the oil can contaminate the bell jar, an alternative method is to place a small white card within the chamber such that a portion of the card is shielded from the source; the area exposed to the source will become progressively darker as film thickness increases, whereas the shielded area remains white.

Check the integrity of the tungsten basket; if any discontinuities are visible, replace the basket. Place ~3-4 cm of gold wire, loosely coiled, into the tungsten basket. Close the bleed-in valve, replace the bell jar carefully on the baseplate gasket, and open the rotary pump–bell jar valve. The bell jar is rough-pumped and the relative vacuum measured with a thermocouple gauge. If the rotary pump is in good condition, the bell jar is clean, and the specimens are not severely outgassing, low vacuum ($10^{-2}$-$10^{-1}$ torr) will be achieved in less than 5 min. Next, the rotary pump–bell jar valve is closed, the rotary pump–diffusion pump valve opened, and the diffusion pump–bell jar valve opened. A discharge gauge is used to monitor high vacuum. Wait until the discharge gauge registers a minimum of $10^{-5}$ torr; recall that higher-vacuum conditions produce finer films.

Next, activate the rotating/tilting stage, turn on the high voltage, and slowly increase the current passing through the tungsten basket until it glows red. In the following step, the filament will become extremely bright, and the intensity is high enough to burn one's eyes. DO NOT LOOK DIRECTLY AT THE FILAMENT. Protect your eyes with cobalt glass or shield the filament and direct your attention to the film-thickness indicator. As the current is slowly increased, the gold will melt at 1063 °C, and a further temperature increase to 1217 °C vaporizes the metal. The card or porcelain plate will become progressively darker as the thin film grows and thickness increases, but will remain white in protected areas. A light gray color corresponds to a film ~50 Å thick, while progressively darker shades correspond to thicker films. An average film thickness of 150-200 Å is suitable for routine coating purposes. When a moderate gold color is observed, turn down the current, turn off the high voltage, and turn off the high-vacuum monitor. If evaporation is continued, a gold reflection color will be observed, and the interior of the bell jar exhibits a gold mirror-like finish. Then close the valve between the bell jar and diffusion pump, open the bleed-in valve, and wait until atmospheric pressure is reached. Remove the specimens, and protect them against heat and humidity until they are examined.

The bell jar interior is thoroughly cleaned with a soft cloth dampened with acetone. Persistent deposits may be removed with polish and the bell jar wiped clean. Check the integrity of the baseplate gasket and its mating surface on the bell jar; gently wipe both surfaces with a lint-free cloth to remove dust. Another evaporation cycle may be performed following the procedure given above, or the system is shut down if it is only irregularly used. The vacuum system shut-down is as follows: replace the bell jar, close the bleed-in valve, open the rotary pump–bell jar valve, and evacuate the system to ~$10^2$ torr. Then close the rotary pump–bell jar valve, open the rotary pump–diffusion pump valve, and turn off the diffusion pump. Wait ~20 min for the diffusion pump to cool, then close the rotary pump–diffusion pump valve, and turn off the rotary pump, main power switch, and cooling water supply.

The vacuum pumps periodically require additional maintenance to remain in optimal working order. The rotary pump oil level, visible through a window, may need replenishing, and the oil should be drained and replaced at least once a year. The diffusion pump oil may require replacement more frequently, particularly in high-humidity environments. When thin-film graininess or irregularity increases, assume that the diffusion pump oil must be replaced. Several water failures and accidental exposure of the hot oil to atmosphere will also deplete the diffusion pump oil; if either of these situations is encountered, change the oil. Different manufacturers of bell jar systems offer procedures for vacuum-pump maintenance; consult the operating manual for a given system.

In the event of a water failure, the diffusion pump oil is protected as follows: immediately turn the diffusion pump off and open the rotary pump–diffusion pump valve. If the diffusion pump was open to the bell jar, close that valve. Simply back-pump the diffusion pump until water can be restored. This procedure will only reduce damage to the oil, not eliminate damage. Power failures are handled by closing all vacuum valves, turning all power switches off, and, if possible, continuing water flow to the diffusion pump. Either a water or a power failure can damage the vacuum system if it is left unattended. Simply be prepared to cope with failures in a logical manner.

## Carbon Thin Films

Carbon films were introduced in transmission electron microscopy as support films (Bradley, 1954), and have been adapted by scanning electron microscopists for two major purposes: a preliminary coating with carbon enhances the adhesion of an evaporated metal coating to the specimen, and carbon films may be used alone to increase the conductivity, though not the secondary electron yield, of nonconductive specimens requiring an X-ray analysis. In the latter situation, carbon is preferred over conventional metal evaporants because carbon is of low atomic weight (12.01), it does not release a detectable X-ray signal, and it does not excessively absorb characteristic X-rays released by the specimen. In comparison, heavy metal coatings absorb low-energy and low-intensity X-rays, as well as emitting characteristic radiation which may mask the true composition of the specimen. Thin films of low-atomic-weight metals such as aluminum may be used for both X-ray analysis and imaging, provided it is understood that sodium, magnesium, and silicon X-rays may be masked or absorbed by the aluminum film. Heinrich (1968) presents mass absorption coefficient tables; reference to these data indicates that some degree of absorption will be encountered with any thin film, but carbon possesses a very low absorption coefficient.

The vacuum bell jar is used in the preparation of carbon thin films, but two spectroscopically pure carbon electrodes are used in place of the tungsten basket. As shown in Fig. 8-3, a pointed electrode is spring-loaded and held in compression against a flat, stationary electrode. The compression between the two electrodes completes the electrical circuit and controls the thickness of the film. Different electrode configurations (e.g., two pointed electrodes) may be used, but are awkward to handle.

Both the bell jar system and samples are prepared as described earlier, with the substitution of carbon planchets and carbon paint for stubs and silver paint. Again, a white card or porcelain plate may be used to estimate film thickness; DeBoer and Brakenhoff (1974) describe a more accurate method which relates film thickness to light absorption. After evacuation to $\sim 10^{-2}$ torr, a carbon arc is formed by passing an alternating current of 20 amp at 30 volts through the electrodes. The carbon will vaporize and form a thin film; thickness is estimated as described earlier.

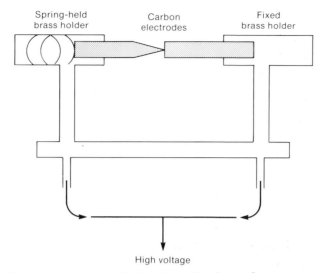

Fig. 8-3. **Arrangement of the electrodes for carbon evaporation.**

Light to moderate gray depositions are desirable. The specimens are then removed and analyzed.

Carbon films may also be used to improve the adhesion between the specimen and a metal thin film. Specimens such as replicas prepared from dental impression media are difficult to coat evenly with evaporated metal; a more even coating results if the specimen is first coated with carbon, then with metal. This sequence may be conducted during a single cycling of the bell jar vacuum system. Most bell jars have multiple high-voltage ports; carbon can be evaporated from one set of electrodes and metal from another set. Each set of electrodes is independently energized, and sequential depositions of carbon and metal are made.

## Film Thickness

Film thickness is defined as either real thickness or mass thickness. The latter refers to the scattering potential of the film and accounts for the variations within ultrathin film structure. The methods used to measure thickness are nontrivial and are more commonly encountered in research situations; these methods are briefly reviewed in order to familiarize the reader with state-of-the art technology. Furthermore, film thickness directly controls resolution of a given "perfect" specimen examined under "perfect" conditions. For example, if 60 Å point-to-point resolution is desired, the film thickness must not exceed 60 Å. High resolution has historically been linked with TEM and STEM, but as the resolution obtainable with the SEM has improved, considerations such as sample preparation become critical.

Flood (1980) categorized thin-film measurement methods into before, during, after, and in situ measurements. Before evaporation, the film thickness can be predicted from

$$T = \frac{W}{4R^2d}$$

where $T$ is thickness, $W$ is amount of evaporant, $R$ is source-target distance, and $d$ is density of evaporant. This calculation crudely estimates film thickness but does not account for incomplete evaporation, alloying between the filament and evaporant, and the source-target geometry.

During evaporation, film thickness can be monitored with a vacuum microbalance (Pearson and Wadsworth, 1965) or a quartz crystal oscillator (Chopra, 1969; Glang, 1970). These sophisticated instruments are effective for measuring continuous films but may provide misleading data for discontinuous films (Peters, 1980).

A variety of methods are used to measure film thickness following deposition. Both the intensity of emitted X-rays (Priyokumos-Singh et al, 1976) and backscattered electrons (Hohn, 1977; Niedrig, 1978) are proportional to film thickness. Alternatively, the real thickness of the film may be measured by the TEM examination of coated latex spheres prepared simultaneously with the specimen (Roli and Flood, 1978; Flood, 1980). Finally, very accurate measurements may be performed using sophisticated double or multiple beam interference methods (Pliskin and Zanin, 1970; Flood, 1980).

## Artifacts of Thermal Evaporation

Various artifacts may be introduced during thermal evaporation. Many of these are usually associated with TEM specimens (Willison and Rowe, 1980), although they may also be encountered by the scanning electron microscopist. Most artifacts arise from improper technique, while others originate from coating a "non-ideal" specimen. For example, it is difficult to evenly coat a very rough surface: the dimensions of prominent features may be exaggerated by their tendency to accumulate excessive evaporant, whereas pits or crevices are shielded and do not acquire a coating. This type of artifact, called *capping*, artificially enhances topographic contrast at the sacrifice of less prominent surface structures.

If the specimen is held stationary during evaporation, artifactual *self-shadowing* may result. Self-shadowing, which is manifested as a line running normal to the shadowing direction, results from shielding by a prominent surface structure. This artifact may be observed if the immobile specimen is coated, repositioned, and recoated, but rarely arises when a rotating/tilting specimen stage is used. Both capping and self-shadowing artifacts are more common with rough-surfaced specimens than with smooth specimens. As will be discussed later in this chapter, very rough surfaces are easily coated using the sputter method. Thermal evaporation is preferred for relatively smooth surfaces and for those specimens requiring a carbon coating for EDS.

Artifacts caused by poor technique include thermal damage and contamination-related problems. *Thermal artifacts* are manifested as very minute melted areas, small holes, or surface discontinuities, and are due to inadequate source-specimen separation (Echlin, 1978). Most thermal artifacts are eliminated by maintaining a 15-cm distance between the source and sample. Highly sensitive specimens may be held on a cold finger, or the evaporation apparatus may be modified by interposing a high-speed rotating aperture between the source and specimen.

*Contamination* can arise before, during, or after evaporation, and produces charging artifacts in the SEM. Before coating, the specimen must be thoroughly cleaned to remove settled dust, hydrocarbons, or any other material which obscures the specimen surface. If the specimen is not cleaned, it is difficult to distinguish the true surface from entrapped dirt. Hydrocarbon contamination may also arise from the vacuum system; vapors may be adsorbed by the specimen and are subsequently entrapped by the thin film. Hydrocarbon films significantly lower the binding capacity of the specimen surface, and consequently the rate of thin-film growth over contaminated versus cleaned areas is different. The different growth rates will produce films of uneven thickness that are poorly adherent. In the SEM, such thin films appear cracked, and in extreme situations, the "film" itself may consist of isolated islands. Because the films are discontinuous, their conductive potential is poor and charging is observed.

In addition to hydrocarbon contaminants, any type of liquid or vapor layer will produce these problems. Recall that the coating of dental impression replicas is difficult; a thin liquid layer of plasticizer at the replica surface inhibits film growth (Echlin, 1981). Contamination is minimized by thoroughly cleaning the bell jar after every procedure, maintaining the vacuum system in optimal working condition, employing cold traps (if the bell jar is so equipped), and stringently cleaning the specimen prior to coating.

Coated specimens are also subject to contamination following evaporation. The specimen can adsorb vapors, which obviously cannot be avoided under normal operating conditions. Settled dust is an easily avoidable problem; always store coated specimens in a sealed container which can itself be stored in a desiccator. Whenever possible, examine these specimens immediately after coating.

## SPUTTER COATING

The most popular thin-film technique in SEM is the sputter coating method. Sputtered and evaporated coatings both serve to increase the secondary electron yield of nonconductive specimens, but the principles of thin-film preparation by these two methods are very different. Whereas resistance heating is based upon the vaporization of a metal at high temperature, sputtering involves the erosion of atoms from a metal target by an energetic plasma. Thermal evaporation is conducted under high-vacuum conditions and is a line-of-sight phenomenon; sputtering is conducted under low-vacuum conditions, and deflection of the liberated metal atoms by residual gas effectively produces a multidirectional source. Because metal atoms approach the specimen from all directions, very rough-surfaced specimens receive a more continuous coating than that obtainable with thermal evaporation.

## Sputter Coaters

Sputtering units may be classified into plasma sputter coaters, radio-frequency sputter coaters (Jackson, 1970), ion beam sputtering units (Clay and Peace, 1981), Penning sputter coaters (Peters, 1980), and magnetron sputter coaters (Nockolds et al, 1982). Plasma sputter coaters are most commonly used in SEM, and may be categorized into diode or triode units, with diode (or direct-current) units most popular. Both of these are discussed below; the more advanced reader should consult the references cited above or the reviews by Echlin (1978, 1981) on other sputtering methods.

The diode (or direct-current, DC) sputter coater is a simple but reliable apparatus (Echlin, 1975a). As shown in Fig. 8-4, it consists of a small bell jar containing a metal target (e.g., gold), which functions as a cathode, and the specimen stage, which functions as an anode. The bell jar is connected to a rotary pump capable of evacuating the system to approximately $10^{-2}$ torr. A leak valve connecting the bell jar to a cylinder of compressed argon or nitrogen, as well as a bleed-in valve to the inert

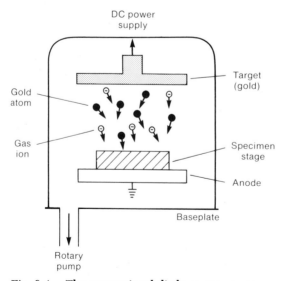

**Fig. 8-4.   The conventional diode sputter coater.**

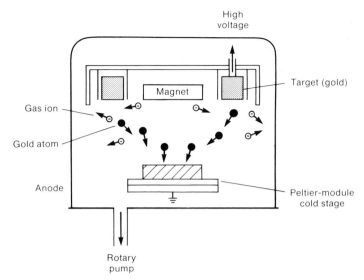

**Fig. 8-5.    The cool diode sputter coater.**

gas or to atmosphere, are also present. A vacuum gauge, high-voltage (synonym: accelerating voltage and high tension) control, current indicator, and timer complete the system.

As will be discussed, the conventional DC sputtering apparatus described above can generate enough heat to damage thermally sensitive specimens. Panayi et al (1977) modified the DC coater in order to eliminate thermal artifacts. A cross section of their cool diode sputter coater is shown in Fig. 8-5; the conventional disk-shaped target has been replaced by an annular ring target. Centered within the annular target is a magnet, and surrounding the target is a pole piece. During operation, this arrangement will deflect electrons created during sputtering outside the zone of the specimen holder. Additional protection against thermal effects is provided by substituting a Peltier-effect cooling module for the conventional specimen stage.

A completely different approach to suppressing thermal damage is to employ a triode sputter coater (Fig. 8-6). Ingram et al (1976) describe a triode apparatus which is a modified DC coater; the typical anode/specimen-cathode/target arrangement is altered by introducing a higher-potential anode. The target is a separate negative electrode, the specimen is at ground potential, and the anode is at high positive potential. The free ions and electrons generated during sputtering will be attracted in the direction of the higher potential difference, which is away from the specimen stage.

## Theory and Method of Sputter Coating

Sputtering basically involves the erosion of target metal atoms (e.g., Au, Ag, Au-Pd, . . .) by an energetic gas plasma (ionized Ar or N) through the creation of a glow discharge. This occurs by a transfer of momentum

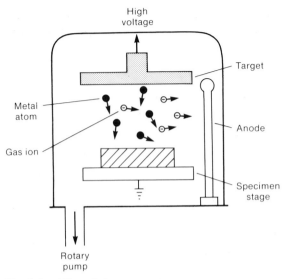

**Fig. 8-6.    The triode sputter coater.**

from the plasma ions to the target metal, which accommodates this excess energy by releasing a metal atom. Wehner and Anderson (1970) cleverly describe sputter coating as three-dimensional billiards, where the cue ball is represented by a plasma ion, and the closely packed metal atoms of the target represent the racked balls. When the argon ion/cue ball strikes the metal atoms/billiard balls, the argon ion transfers its momentum to the target. The target reacts by transmitting this excess energy to surrounding atoms in a direction opposite that of the initial encounter. As this wave of momentum reaches the surface of the target, a metal atom is ejected and attracted by the specimen, where a thin film will be accreted. Heat is a secondary by-product of this encounter; the target will dissipate some of the energy transferred by the plasma ions as heat. To avoid radiant heat damage, the target and specimens are separated by a gap of ~2-5 cm.

The general trajectory of the ejected metal atoms is toward the anode, but because sputtering is conducted under low vacuum, the presence of residual gas causes a deflection of the metal atoms, which then arrive at the specimen surface from many directions. It is also probable that free metal atoms collide with one another while in free space, coalesce to form a nucleus, and then are intercepted by the specimen surface (Renou and Gillet, 1978). The multidirectional nature of sputtering produces a coating in areas which are hidden from a line of sight of the target, meaning that this method is most effective for coating rough topographies.

A purple glow discharge is visible during sputtering, and arises from the interaction of electrons ejected from the target with free gas molecules, producing an ion and free electron. In the conventional DC sputter coater, these electrons may strike the anodic specimen, which consequently may undergo a temperature rise of roughly 50-76 °C above ambient (Echlin, 1978).* When thermal effects were recognized, the cool diode and triode sputter coaters were developed, and the source of heat (i.e., free electrons) was eliminated. In the cool DC unit, electrons are deflected by magnetic lines of flux away from the specimen stage, which does not exceed temperatures of 32-37 °C (Panayi et al, 1977). The triode sputter coater diminishes electron bombardment by attracting free electrons toward the high-potential anode, and limits specimen temperatures to less than 54 °C (Panayi et al, 1977). Temperature can be minimized in the conventional DC sputter coater by cooling the specimen stage with water and by carefully limiting the accelerating voltage and emission current during sputtering.

In addition to potential damage to thermal-sensitive specimens, heating affects the thin-film granularity. Sputtered thin films grow by the same process as do evaporated films; recall from the "Theory of Evaporation" that lowering the specimen temperature enhances the fineness of the thin film. Low temperature reduces the surface mobility of the metal atoms and produces smaller and more uniformly sized nuclei per unit area (Echlin and Kaye, 1979; Echlin et al, 1980, 1982; Robards et al, 1981). During a later stage of film growth, lower temperatures favor the formation of many small islands of condensed metal, which in turn reduces film granularity. While these effects are most pronounced with high-resolution SEM, they must not be ignored during routine analyses.

These factors indicate that the cool diode and triode sputter coaters have definite advantages for SEM. Owners of conventional DC sputtering units should not be discouraged; by carefully controlling the accelerating voltage and plasma current, finer films free from thermal artifacts can be produced.

The deposition rate is dependent on the characteristics of the plasma and the target metal and on the operating parameters of the sputter coater. The gases used for sputtering are argon (atomic number 18) or nitrogen (atomic number 7), and have been selected because their effective diameter is relatively large. Smaller atoms such as hydrogen are ineffective because they cannot carry sufficient momentum to affect noble metals. Echlin (1978) compared the effect of a hydrogen plasma ion bombarding a metal target to that of a marble striking a bowling ball. With a more massive gas, the effective diameters of the plasma ions and the metal target are more similar, and the probability of the ejection of a metal atom increases. Sputtering with argon is more rapid than sputtering with nitrogen because of the greater mass of argon.

---

*Temperature measurements at the specimen level are usually made with thin-film thermocouples. Clark et al (1976) is an interesting paper which discusses temperature measurement using this method.

Although compressed-gas cylinders of argon or nitrogen are not prohibitively expensive, many microscopists use air to form the plasma. This is a very poor choice for several reasons: sputtering is less efficient with gases of low atomic weight, the deposition rate is decreased, water vapor and/or particles will contaminate the specimen, and the resulting thin films are of poor quality. As will be discussed, uncontrolled contamination is a severe problem in sputter coating, and sputtering with air normally produces obvious contamination-related artifacts. Although sputtering itself is a low-vacuum phenomenon, the specimen chamber is backfilled until a pure nitrogen or argon atmosphere is produced, which eliminates potential contaminants. In short, always use nitrogen or argon for sputter coating.

The target metals used for sputtering include the noble metals and their alloys, as well as other metallic elements such as nickel, chromium, copper, silver, aluminum, tantalum, and tungsten. This is a distinct advantage that sputtering has over resistance heating: whereas thermal evaporation is restricted to nonrefractory metals, almost any metal regardless of its melting point can be sputtered. For example, the refractory metals tantalum and tungsten can be sputtered but cannot be thermally evaporated (Adachi et al, 1976). The refractory metals are appropriate for high-resolution SEM, whereas pure gold or gold-palladium alloys are used for routine SEM. The influence of the melting point of the target metal, as discussed under "Theory of Evaporation," holds true for sputtering; metals having higher melting points produce more amorphous (less grainy) thin films (Broers and Spiller, 1980). Echlin and Kaye (1970) systematically compared the appearance of different metallic thin films prepared under various operating conditions; their micrographs vividly support this thesis (Fig. 8-7). Further, film granularity may be suppressed by sputtering, for example, a gold-palladium alloy rather than pure gold. The manufacturers of sputter coaters have available an extensive selection of metal targets; a gold or gold-alloy target is typically supplied with the unit.

The operation of the sputter coater also influences the appearance of the thin films. These factors include the accelerating voltage (roughly 1.5-2.5 keV) discharge current ($\sim$20 mA), relative vacuum, and temperature of the specimen. These interrelated factors are more clearly understood in the context of the operation of the sputter coater, which is presented below.

The first stage of sputter coating involves specimen mounting, positioning in the sample chamber, and cooling of the specimen stage (where appropriate). Sputter coaters equipped with a water-cooled or Peltier-module cold stage require approximately 10 min for the specimen to cool from ambient temperatures to 0-5 °C. If the rotary pump is equipped with a cold trap, it should be filled with liquid nitrogen. The specimen

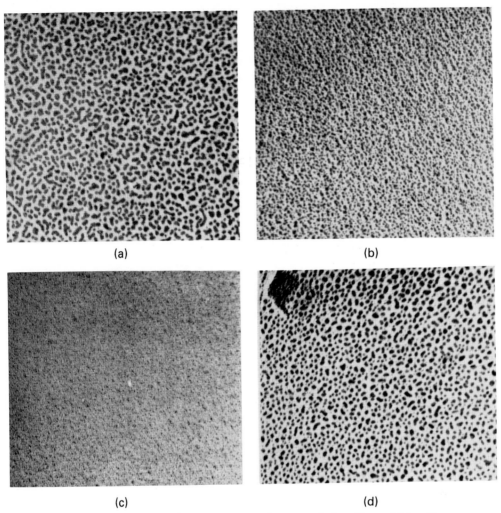

(a)

(b)

(c)

(d)

Fig. 8-7. The appearance of sputtered thin films of (a) gold, (b) gold-palladium, and (c) platinum. An evaporated gold film is shown in (d). (Courtesy of Dr. Patrick Echlin)

chamber is then evacuated to $\sim 10^{-1}$ torr, flushed at least three times with pure, anhydrous argon or nitrogen, and then re-evacuated to $10^{-1}$-$10^{-2}$ torr. Flushing the vacuum chamber ensures that it is free from residual vapors and gases and establishes a pure argon or nitrogen atmosphere. The specimen and adhesives will outgas, and both represent potential contamination sources: make sure that the specimen is clean and dry before initiating the coating procedure. Also, the chamber should not be pumped indefinitely, because the probability of backflow from the rotary pump increases as the ultimate pressure of the pump is reached. Backstreaming can be avoided by a slight but continuous flow of argon or nitrogen through the specimen chamber. The only other means to eliminate backflow is to employ cryosorption liquid-nitrogen pumps rather than rotary pumps; such a system is described by deHarven et al (1978).

The desired film thickness $T$ Å can be calculated from the following equation (Echlin, 1975a):

$$T = C \cdot V \cdot t \cdot k$$

where $C$ is the plasma discharge in mA, $V$ is the accelerating voltage in keV, $t$ is time in minutes, and $k$ is a gas constant. For argon/gold, $k = 5$; for nitrogen/gold, $k = 2$.

This equation may be used to precalculate the desired film thickness ($\sim$150-250 Å) by varying time. The actual sputtering procedure is initiated (at $10^{-2}$-$10^{-1}$ torr and $\sim$40 °C) by turning on the high voltage (synonym: high tension) to 1.8-2.5 keV and slowly opening the gas leak valve until a discharge current of 20 mA and a vacuum of $\sim$$10^{-1}$ torr are achieved. The desired length of time ($\sim$3-5 min) is then entered into the coater timer, and sputtering is carried out for this duration. If nitrogen is used as the plasma, increase the duration of coating because the value of $k$ for nitrogen/gold is 2. Though the above calculation is most valid for smooth surfaces, it provides an adequate description of coating thickness for rough surfaces. After coating, the high voltage is turned off, the rotary pump is turned off, and the chamber is immediately backfilled with argon until atmospheric pressure is reached. Turning off the rotary pump and backfilling are performed as one motion to prevent backstreaming of pump oil into the specimen chamber. Some sputter coaters have a bleed-in valve to atmospheric air; this may be used, provided that care is taken to avoid excessive humidity and/or particle contamination. The specimens are then removed and promptly examined, or may be stored in a dust-free container in a desiccator.

Finally, the specimen chamber is cleaned with a cloth dampened with acetone. Unfortunately, many microscopists do not clean the sputter coater as faithfully as they clean the bell jar system; contamination and the successive layers of sputtered films will degrade both the vacuum and the thin-film appearance. Water vapor evolved from the specimen or from contaminated gas cylinders also adversely affects sputtering; during sputtering, the presence of residual water vapor is revealed as a blue tint in the purple discharge. These contamination sources also affect the target, as will handling during target changing. Target contamination is manifested as degradation of vacuum when the high voltage is turned on. The target is carefully cleaned by gently wiping with a lint-free cloth, followed by actual sputtering without specimens (Echlin, 1978).

In summary, good thin films exhibit minimal granularity and are obtained by sputtering in a clean system (specimens, vacuum chamber, target, and anhydrous inert gas) at a relatively low deposition rate (low accelerating voltage and discharge current). Do not prepare excessively thick ($>$300 Å) coatings; fine surface features will be masked.

## Artifacts of Sputter Coating

Most sputtering artifacts arise from either contamination or thermal damage. *Contamination* arises from outgassing of the specimen, hydrocarbons evolved from the rotary pump, and contaminated gas cylinders (Maissel, 1970). Specimen outgassing is minimized by thorough cleaning and drying prior to coating; Echlin (1978) recommended heating at ~40 °C overnight to ensure that porous specimens dry completely. Hydrocarbon contamination is minimized by using rotary pump oils having a low backstreaming coefficient, high-capacity pumps with alumina traps in the backing line, and high-purity gas that is passed through a filter (Echlin, 1981). Artifacts resulting from contamination are manifested either as cracks and irregularities in the coating or by an abnormal coloring (e.g., red, blue, or green as opposed to gold) of the sputtered film (Simmens, 1975; Echlin et al, 1980). If any of these problems arises, the microscopist should trace and eliminate the source of contamination.

*Thermal damage* was recognized shortly after sputtering was first used in SEM but has been eliminated with the advent of cool sputter coaters. Heat-sensitive specimens such as delicate biological samples (Rowsowski and Glider, 1977; Braten, 1978) and low-melting-point hydrocarbon crystals (Ingram et al, 1976; Panayi et al, 1977) have been used to evaluate thermal damage. Thermal artifacts are manifested as pitting or melting of the film and are observed at excessive accelerating voltages and/or plasma currents. If the specimen is sensitive to temperatures of roughly 50-75 °C above ambient, care must be taken during both coating and SEM examination: heating is a natural by-product of electron-beam irradiation (Munger, 1977).

Other artifacts that may be encountered with sputter coating include surface etching of the specimen by stray electrons (in conventional DC coaters) and bombardment of the specimen by high-energy metal atoms. Holland (1976) described *decoration artifacts,* or imperfections in the film, which probably originated from contamination or thermal damage. A more common problem is the irregular film coating that may be observed on extremely rough surfaces, such as those in specimens having deep, irregular pits. Slobodrian et al (1978) described a method for coating such specimens, but usually the microscopist simply tolerates such problems by altering the SEM operating parameters.

# COMPARISON OF COATING METHODS

The microscopist who has access to both a sputter coater and a thermal evaporator will choose one method or the other on the basis of the characteristics of the specimen. Rough-surfaced specimens will receive a more continuous coating of sputtered metal because of the multi-directional nature of sputtering (DeNee and Walker, 1975). Smoother surfaces are better coated using rotary evaporation. Specimens requiring

X-ray analysis, regardless of their surface structure, are coated with a carbon thin film with the bell jar apparatus. Shadowing, which may be desirable to enhance the fidelity of microstructural features, presupposes that the specimen is naturally conductive or precoated (e.g., with carbon). Specimens which resist film adhesion during metal evaporation (e.g., some dental impression replicas) should be precoated with a thin carbon film before the evaporated metal film is applied; alternatively, sputter coat the replica, because film adhesion is not a problem with sputtering (Munger, 1977).

Thin-film granularity is suppressed by several factors. Always clean the specimen and both coaters thoroughly to prevent outgassing. Cooling the specimen to decrease the rate of thin-film growth also suppresses granularity. For moderate-resolution SEM, gold or gold-palladium thin films are acceptable, but higher-resolution examinations (>100 Å) require coating with metals/alloys having higher melting points. For evaporation, this would include platinum and other nonrefractory metals; with sputtering, these same metals or refractories (e.g., tungsten or tantalum) may be prepared as thin films.

Film thickness should be minimal; excessively thick films will mask fine structural features, while discontinuous films decrease the secondary-electron yield. Further, the calculations for film thickness presented earlier assume that the target is flat and regular. Consequently, the exact film thickness for irregular surfaces as calculated is typically invalid. These calculations are useful, however, as crude estimations of thickness. For critical applications it may be necessary to employ one of the more rigorous measuring methods.

Coated specimens are examined at moderate to high voltages (10-20 keV) with a small spot size. Recall from Chapter 1 the parameters governing SEM resolution: high voltages and very large spot sizes increase the depth of penetration. Thus, the beam may pass through the film and excite the underlying specimen, which may be a problem with heat-sensitive specimens. Always use the optimal accelerating voltage and spot size when examining these specimens. Finally, after examining coated specimens, it may be necessary to remove the gold coating. Sela and Boyde (1977) describe cyanide removal of gold thin films. Usually it is easiest to postpone SEM examinations until all other testing has been completed.

# REFERENCES

Abermann, R., and M. M. Salpeter (1974) Visualization of desoxyribonucleic acid by protein film adsorption and tantalum-tungsten shadowing. *J. Histochem. Cytochem* 22:845.

Adachi, K., et al (1976) High resolution shadowing for electron microscopy by sputter deposition. *Ultramicroscopy* 2:17.

Baumeister, W., and M. Hahn (1978) Specimen supports. In: *Principles and Techniques of Electron Microscopy*, vol. 8. (Hayat, M.A., ed.) Van Nostrand Reinhold, New York, p 1.

Blaschke, R. (1980) Three examples of coating artifacts in scanning electron microscopy. *Proc. Roy. Soc. Micros.* 15:280.

Bradley, D. E. (1954) Evaporated carbon films for use in electron microscopy. *Brit. J. Appl. Phys.* 5:65.

_____ (1958a) A new approach to the problem of high resolution shadow casting: The simultaneous evaporation of platinum and carbon. *Proc. 4th Int. Cong. EM* 1:428.

_____ (1958b) Simultaneous evaporation of platinum and carbon for possible use in high resolution shadow casting for electron microscopy. *Nature* 181:875.

_____ (1959) High-resolution shadow-casting techniques for the electron microscope using the simultaneous evaporation of platinum and carbon. *Brit. J. Appl. Phys.* 10:198.

_____ (1967) Replica and shadowing techniques. In: *Techniques for Electron Microscopy*, 2d ed. (Kay, D. H., ed.) Blackwell Sci. Pub., Oxford, p 96.

Braten, T. (1978) High resolution scanning electron microscopy in biology: Artifacts caused by the nature and mode of application of the coating material. *J. Micros.* 113:53.

Broers, A. N., and E. Spiller (1980) A comparison of high resolution scanning electron micrographs of metal film coatings with soft X-ray interference measurements of the film roughness. *SEM, Inc.* 1:201.

Chopra, K. L. (1969) *Thin Film Phenomena*. McGraw-Hill, New York.

Clark, J., et al (1976) Thin film thermocouples for use in scanning electron microscopy. *IITRI/SEM* 1:83.

Clay, C. S., and G. W. Peace (1981) Ion beam sputtering: An improved method of metal coating SEM samples and shadowing CTEM samples. *J. Micros.* 123:25.

DeBoer, J., and G. J. Brakenhoff (1974) A simple method for carbon film thickness determination. *J. Ultrastr. Res.* 49:224.

deHarven, E., et al (1978) Sputter coating in oil contamination-free vacuum for scanning electron microscopy. *SEM, Inc.* 1:167.

DeNee, P. B., and E. R. Walker (1975) Specimen coating techniques for SEM—a comparative study. *IITRI/SEM*, p 225.

Echlin, P. (1975a) Sputter coating techniques for SEM. *IITRI/SEM*, p. 217.

_____ (1975b) Contamination in the SEM. *IITRI/SEM*, p 679.

*_____ (1978) Coating techniques for scanning electron microscopy and X-ray microanalysis. *SEM, Inc.* 1:109.

_____ (1981) Recent advances in specimen coating techniques. *SEM, Inc.* 1:79.

*Echlin, P., and G. Kaye (1979) Thin films for high resolution conventional SEM. *SEM, Inc.* 2:21.

Echlin, P., et al (1980) Improved resolution of sputter coated films. *SEM, Inc.* 1:163.

_____ (1982) Low voltage sputter coating. *SEM, Inc.* 1:29.

Filshie, B. K., and C. D. Beaton (1980) 3-D structure of thin metal coatings at high resolution in the SEM. *Proc. 7th Eur. Conf. EM* (Leiden, The Netherlands) 2:792.

*Flood, P. R. (1980) Thin film thickness measurement. *SEM, Inc.* 1:183.

Glang, R. (1970) Vacuum evaporation. In: *Handbook of Thin Film Technology*. (Maissel, L. I., and R. Glang, eds.) McGraw-Hill, New York, ch 7.

Hagler, H. F., et al (1977) A simple electron beam gun for platinum evaporation. *J. Micros.* 110:149.

Harris, W. J. (1975) A universal metal and carbon evaporation accessory for electron microscope techniques and a method for obtaining repeatable evaporation of platinum-carbon. *J. Micros.* 105:265.

Heinrich, K. F. J., ed. (1968) *Quantitative Electron Probe Microanalysis.* NBS Pub. 298. U.S. Govt. Printing Office, Washington, DC, p 123.

Henderson, W. J., and K. Griffiths (1972) Shadow casting and replication. In: *Principles and Techniques of Electron Microscopy,* vol. 2. (Hayat, M.A., ed.) Van Nostrand Reinhold, New York, p 151.

Hodgkin, N. M., and L. E. Murr (1974) Quantitative study of vapour-deposited metal coatings for SEM. *Micro Sci.* 2:129.

Hohn, F. T. (1977) Angular dependence of electron intensities backscattered by carbon films. *Optik* 47:491.

Holland, L. (1970) *Vacuum Deposition of Thin Films.* Chapman and Hall, Ltd., London.

Holland, V. F. (1976) Some artifacts associated with sputter coated samples observed at high magnification in the scanning electron microscope. *IITRI/SEM* 1:71.

Ingram, P., et al (1976) Some comparisons of the techniques of sputter (coating) and evaporative coating for scanning electron microscopy. *IITRI/SEM* 1:75.

Jackson, G. N. (1970) Radiofrequency sputtering. *Thin Solid Films* 5:209.

Johansen, B. V. (1974) Brightfield electron microscopy of biological specimens. II. Preparation of ultra-thin carbon support films. *Micron* 5:209.

Maissel, L. I. (1970) Application of sputtering to the deposition of films. In: *Handbook of Thin Film Technology.* (Maissel, L. I., and R. Glang, eds.) McGraw-Hill, New York, ch 4.

Molcik, M. (1967) Technique to reduce the grain size of shadowing alloys used in electron microscopy. *Prak. Metallog.* 4:628.

Morris, J. E., and T. J. Coutts (1977) Electrical conductance in discontinuous metal films: A discussion. *Thin Solid Films* 47:1.

*Munger, B. L. (1977) The problem of specimen conductivity in scanning electron microscopy. *IITRI/SEM* 1:481.

Nagatani, T., and M. Saito (1974) Structure analysis of evaporated films by means of TEM and SEM. *IITRI/SEM,* p 51.

Neugebauer, C. A. (1970) Condensation, nucleation, and growth of thin films. In: *Handbook of Thin Film Technology.* (Maissel, L. I., and R. Glang, eds.) McGraw-Hill, New York, ch 8.

Niedrig, H. (1978) Backscatter electrons as a tool for film thickness determination. *SEM, Inc.* 1:841.

Nockolds, C. E., et al (1982) Design and operation of a high efficiency magnetron sputter coater. *SEM, Inc.* 3:907.

Panayi, P. N., et al (1977) A cool sputtering system for coating heat-sensitive specimens. *IITRI/SEM* 1:463.

Pearson, S., and N. J. Wadsworth (1965) A robust torsion balance which can detect a force of $2 \times 10^{-8}$ dyne. *J. Sci. Inst.* 42:150.

Pease, R. W., and J. F. Bailey (1975) Thin polymer films as non-charging surfaces for scanning electron microscopy. *J. Micros.* 104:281.

Peters, K. R. (1980) Penning sputtering of ultrathin metal films for high resolution electron microscopy. *SEM, Inc.* 1:143.

Pfefferkorn, G. E. (1973) Techniques for non-conducting samples. *IITRI/SEM,* p 751.

Pliskin, W. A., and S. J. Zanin (1970) Film thickness and composition. In: *Handbook of Thin Film Technology*. (Maissel, L. I., and R. Glang, eds.) McGraw-Hill, New York, ch 11.

Priyokumas-Singh, S., et al (1979) Thickness measurement of single and composite thin metal films using the X-ray fluorescence technique. *Thin Solid Films* 59:51.

Renou, A., and M. Gillet (1978) Formation of gold particles in a flowing argon system: Electron microscopy of the density, size distribution, and size dispersion. *J. Crystal Growth* 44:190.

Robards, A. W., et al (1981) Specimen heating during sputter coating. *J. Micros.* 124:143.

*Roli, J., and P. R. Flood (1978) A simple method for the determination of thickness and grain size of deposited films as used in non-conductive specimens for SEM. *J. Micros.* 112:359.

Rowsowski, J. R., and W. V. Glider (1977) Comparative effects of metal coating by sputtering and by vacuum evaporation on delicate features of euglenoid flagellates. *IITRI/SEM* 1:471.

Sela, J., and A. Boyde (1977) Cyanide removal of gold from SEM specimens. *J. Micros.* 111:229.

Shaffner, T. J., and J. W. S. Hearle (1976) Recent advances in understanding specimen charging. *IITRI/SEM* 1:61.

Shiflett, C. C. (1968) Evaporated films. In: *Thin Film Technology*. (Berry, R. W., et al, eds.) Van Nostrand Reinhold, New York, p 113.

Sikorsky, J., et al (1968) A new preparative technique for examining polymers in the SEM. *J. Sci. Inst.* 1:29.

Simmens, S. C. (1975) An observation of the metallizing of specimens for scanning electron microscopy using cathode sputtering. *J. Micros.* 105:233.

Slayter, H. S. (1976) High resolution replicating of macromolecules. *Ultramicroscopy* 1:341.

_____ (1978) Fine features of glycoproteins by high resolution replication. In: *Principles and Techniques of Electron Microscopy*, vol 9. (Hayat, M.A., ed.) Van Nostrand Reinhold, New York, p 175.

_____ (1980) High resolution metal coatings of biopolymers. *SEM, Inc.* 1:171.

Slobodrian, M. L., et al (1978) Metallic deposition in specimens presenting cavities using the sputter coater. *J. Micros* 112:365.

Smith, P. R., and I. E. Ivanov (1980) Surface reliefs computed from micrographs of isolated heavy metal shadowed particles. *J. Ultrastr. Res.* 71:25.

Smith, P. R., and J. Kistler (1977) Surface reliefs computed from micrographs of heavy metal shadowed specimens. *J. Ultrastr. Res.* 61:124.

Wehner, G. K., and G. S. Anderson (1970) The nature of physical sputtering. In: *Handbook of Thin Film Technology*. (Maissel, L. I., and R. Glang, eds.) McGraw-Hill, New York, ch 3.

Williams, R. C., and R. W. G. Wyckoff (1946) Applications of metallic shadow casting to microscopy. *J. Appl. Phys.* 17:23.

Willison, J. H. M., and A. J. Rowe (1980) Replica, shadowing, and freeze-etching technique. In: *Practical Methods in Electron Microscopy*, vol 8. (Glauert, A. M., ed.) American Elsevier, New York.

Zingsheim, H. P., et al (1970) Shadow casting and heat damage. *Proc. 7th Int. Cong. EM* 1:411.

*Recommended reading.

# Appendix A:
# Glossary of SEM Terminology

**ABSORPTION.** In SEM, refers to absorption by the specimen of X-rays originating from deep within the excitation volume, resulting in ionization of the
absorbing atom and reduction of the FLUORESCENT YIELD.

**ABSORPTION EDGE ($K_{ab}$).** A unique energy manifested as an abrupt discon-
tinuity in an X-ray spectrum related to both characteristic X-ray energy/
wavelength and accelerating voltage.

**ACCELERATING VOLTAGE.** The difference in potential between the filament (cathode) and the anode, causing acceleration of the electrons by 2 to
30 keV. See also DEPTH OF PENETRATION; RESOLUTION.

**ADC.** See ANALOG-TO-DIGITAL CONVERTER.

**AMPLIFIER.** A component of the energy-dispersive spectrometer responsible
for shaping and magnifying the X-ray signal, which enhances the SIGNAL-
TO-NOISE RATIO.

**ANAGLYPHS.** Stereo pairs observed by passing the right image through a red
filter and the left image through a green filter; the three-dimensional effect
is perceived by wearing red-green lenses. Compare with POLARIZED
STEREO PROJECTION.

**ANALOG-TO-DIGITAL CONVERTER (ADC).** A device that converts a varying signal (e.g., voltage) into numerical values suitable for analysis by a digital
computer.

**ANODE.** A component of the electron gun that has a large positive potential
relative to the FILAMENT.

**APERTURES.** Molybdenum or platinum strips (multiple apertures) or disks
(single apertures) having minute openings 30 to 200 $\mu$m in diameter which
intercept stray electrons from the imaging beam.

**ASTIGMATISM.** An optical aberration manifested as a focus defect in which
electrons in different axial planes focus at different points and caused
by minute inhomogeneities in the magnetic lens coiling. See also
STIGMATORS.

**ATOMIC NUMBER IMAGING.** A technique for revealing compositional and
topographic information about a specimen based on its yield of backscattered
electrons; that yield is a function of atomic number and is manifested as
different levels of contrast. Synonymous with backscattered electron imaging
and atomic number contrast.

**AUGER ELECTRON.** One of the electrons emitted as a result of transfer of
energy from the usual X-ray photon to an orbital electron during irradiation
of low-atomic-weight materials. Auger electrons have a characteristic energy
detected as peaks in the energy spectra of the secondary electrons generated.

**BACKGROUND.** Any degradation of an image or spectrum caused by a reduction in the SIGNAL-TO-NOISE RATIO.

**BACKGROUND SUBTRACTION.** A method of X-ray spectrum manipulation that removes the complex X-ray continuum from the characteristic X-ray peaks, usually in preparation for quantitative analysis. See also CONTINUUM.

**BACKSCATTER COEFFICIENT ($\eta$).** The fraction of beam electrons which escape from a specimen, a function of atomic number.

**BACKSCATTER CROSS SECTION ($\sigma$).** The probability of an elastic scattering event.

**BACKSCATTERED ELECTRON DETECTOR.** A solid-state detector, its design based upon the cosine distribution of BSE around the primary beam. See also ROBINSON DETECTOR.

**BACKSCATTERED ELECTRONS (BSE).** An information signal arising from elastic (electron-nucleus) collisions, wherein the incident electron rebounds from the interaction with a small (~20%) energy loss. The BSE yield is strongly dependent upon atomic number, qualitatively describes the origin of characteristic X-rays, and reveals both compositional and topographic information about the specimen. See also ATOMIC NUMBER IMAGING.

**BEAM CURRENT.** A measure of the number of imaging electrons incident upon a specimen in a unit of time.

**BEAM DIAMETER.** The width of the electron beam. Synonymous with SPOT SIZE.

**BEER'S LAW.** A description of X-ray absorption relating transmitted intensity to the mass absorption coefficient and material density.

**BENCE-ALBEE METHOD.** An empirical, quantitative EDS correction for mineralogical specimens.

**BERYLLIUM WINDOW.** A very thin (~7.5 $\mu$m thick), relatively X-ray–transparent window separating the X-ray detector from the vacuum chamber, and serving to protect the detector from damage; cf. WINDOWLESS DETECTOR.

**BIAS.** In the SEM, the potential difference between the filament and grid cap which controls the beam current; in electronics, a voltage applied to an electronic device (e.g., a solid-state detector) to determine the portion of the characteristic of the device at which it operates.

**BINDING ENERGY.** The energy holding an orbital electron in a nuclear field; the binding energy must be exceeded for the emission of characteristic X-rays.

**BLACK-LEVEL SUBTRACTION.** A method of signal processing that relies upon differential amplification to enhance the number of gray levels in an image.

**BOHR MODEL.** A simplified model of the atom defining the location of concentric electron shells relative to the nucleus.

**BRAKING RADIATION.** See CONTINUUM.

**BREMSSTRAHLUNG.** See CONTINUUM.

**BSE.** See BACKSCATTERED ELECTRONS.


**CAPPING.** A thin-film artifact caused by the excessive deposition of metal at prominent surface features and manifested as an exaggeration of the dimensions of those features.

**CARBON FILMS.** Thin films applied to nonconductive surfaces which enhance conductivity without severely limiting the emission of X-rays. Also, specimens which resist coating by evaporated metals may be precoated with a carbon film, which enhances the ability to capture subsequently evaporated metal atoms.

**CATHODE-RAY TUBE.** An electronic tube that permits the visual display of electronic signals.

**CATHODOLUMINESCENCE (CL).** A radiative transition wherein low-energy light photons are released during electron irradiation.

**CATHODOLUMINESCENCE DETECTOR.** Ellipsoidal lenses or mirrors positioned around the specimen which collect light emitted from the specimen during irradiation.

**CELLULOSE ACETATE TAPE.** A replicating medium used to form negative impressions and to clean or protect fracture surfaces.

**CHARACTERISTIC X-RAYS.** X-rays of unique energy and wavelength released during irradiation by the atoms of a specimen. When detected and analyzed, characteristic X-rays reveal the elemental composition of the specimen.

**CHARGING.** An imaging artifact manifested as bright, horizontal stripes across the visual CRT or print and resulting from the accumulation of electrons by a nonconductive specimen or contaminants.

**CHROMATIC ABERRATION.** Variation in the focal length of a lens as a function of the wavelength of the radiation passing through it, caused by fluctuations in the high voltage.

**CL.** See CATHODOLUMINESCENCE.

**CLIFF AND LORIMER METHOD.** A quantitative EDS correction used for the analysis of thin sections.

**COLD FINGER.** A liquid-nitrogen–cooled cold trap used to reduce contamination levels in vacuum chambers.

**COLUMN LINER TUBE.** A hollow metal tube extending from the anode to the final aperture which prevents contamination of the SEM imaging system components.

**CONDENSER LENS.** A demagnifying lens (or series of lenses) located immediately beneath the anode that reduces the diameter of the electron beam.

**CONTINUUM.** The noncharacteristic X-rays released by an irradiated specimen and caused by deceleration of the incident electrons by interaction with a nuclear field. Synonymous with braking radiation, Bremsstrahlung, and white radiation.

**CONTRAST.** The ratio of black to white in an image.

**COOL DIODE SPUTTER COATER.** An apparatus which eliminates thermal damage of the specimen by deflecting electrons away from the level of the specimen.

**COUNT RATE.** The rate at which the components of the EDS system sort and count X-ray signals. See also DEAD TIME.

**CRITICAL EXCITATION EDGE ENERGY.** The minimum energy required for excitation of an X-ray line, usually two to three times greater than the X-ray energy.

**CRT.** See CATHODE-RAY TUBE.

**DC SPUTTER COATER.** See DIODE SPUTTER COATER.

**DEAD TIME.** The total time during which the X-ray spectrometer is processing information and is unavailable to accept input data.

**DECORATION ARTIFACTS.** Imperfections introduced into thin films by the native structure of the evaporated or sputtered metal.

**DEFLECTION YOKE.** A component of the imaging system responsible for moving the electron beam in a raster pattern across the specimen, and located either within or near the final lens.

**DEMAGNIFYING LENSES.** Convergent lenses which reduce the diameter of the electron beam.

**DENTAL IMPRESSION MEDIA.** Polysulfide- or silicone-based materials used in the preparation of large-scale negative REPLICAS.

**DEPTH OF FIELD.** The ability of an imaging system to maintain focus in a field of view despite surface irregularities.

**DEPTH OF PENETRATION.** The distance the electron beam penetrates beneath the surface of a sample, which depends primarily upon specimen density and accelerating voltage. See also EXCITATION VOLUME.

**DIFFUSION PUMP.** A vacuum pump capable of producing the high-vacuum ($\geq 10^{-4}$ torr) condition.

**DIODE SPUTTER COATER.** An apparatus for producing sputtered thin films wherein the metal target functions as a cathode and the specimen as the anode. Synonymous with direct current sputter coater. See also COOL DIODE SPUTTER COATER.

**DIRECT CURRENT SPUTTER COATER.** See DIODE SPUTTER COATER.

**DISCHARGE CURRENT.** In sputtering, the passage of an electric current (measured in milliamperes) through a flowing gas to generate the plasma.

**DOT MAP.** The imaging of the sites of X-ray emission from a specimen. Synonymous with X-RAY MAP.

**EBIC.** See ELECTRON BEAM INDUCED CURRENT.

**EDS.** See ENERGY-DISPERSIVE SPECTROSCOPY.

**ELASTIC COLLISIONS.** Electron-nucleus interactions which produce BACKSCATTERED ELECTRONS; cf. INELASTIC COLLISIONS.

**ELECTROMAGNETIC FOCUS.** Convergence of an electron beam by magnetic fields produced by the passage of a current through a magnetic lens.

**ELECTRON BEAM INDUCED CURRENT (EBIC).** A method for detecting silicon damage in semiconductors based upon the production of electron-hole pairs by the irradiating beam.

**ELECTRON GUN.** Collectively, the filament, shield, and anode, which are connected to the high-voltage supply of the SEM and produce the imaging electron beam. See also FIELD EMISSION SOURCE; LANTHANUM HEXABORIDE SOURCE.

**ELECTRON-GUN EVAPORATION.** A method of producing thin films of refractory metals wherein the evaporant is the anode target and is heated by radiation from a cathode maintained at 2.0 to 2.5 keV.

**ELECTRON-HOLE PAIRS.** Filled valence band and empty conduction band electrons in an electronic energy level.

**ELECTRON SIGNALS.** See AUGER ELECTRON; BACKSCATTERED ELECTRONS; SECONDARY ELECTRONS.

**ELECTROSTATIC FOCUS.** Convergence of a beam of electrons by an electrostatic field generated between two electrodes at different potentials; e.g., the ELECTRON GUN is focused electrostatically.

**EMPTY MAGNIFICATION.** The level at which increased magnification does not improve resolution.

**END CAP.** Housing of the cold finger/detector assembly in the X-ray detector.

**ENERGY-DISPERSIVE SPECTROSCOPY (EDS).** A method of X-ray analysis which discriminates among the energy levels of characteristic X-rays produced during electron-beam irradiation. Synonym: energy dispersive X-ray analysis. Compare with WAVELENGTH-DISPERSIVE SPECTROSCOPY.

**ESCAPE PEAK.** An artifact observed in an X-ray analysis; manifested as a peak at energy 1.74 keV (the silicon K$\alpha$ peak) less than the major line detected, and due to excitation of the Si(Li) detector at too low an accelerating voltage.

**E-T DETECTOR.** See EVERHART-THORNLEY DETECTOR.

**EVAPORATED THIN FILMS.** See ELECTRON-GUN EVAPORATION; THERMAL EVAPORATION.

**EVERHART-THORNLEY DETECTOR.** A secondary electron detector that employs a positively charged Faraday cage to attract low-energy secondary electrons toward the scintillator.

**EXCITATION VOLUME.** The volume beneath the specimen surface where data signals originate as the electron beam spreads laterally and secondary ionizations occur. Synonymous with information depth.

**EXTRACTION REPLICAS.** Impressions of a surface which remove and preserve inclusions, oxides, sulfides, etc.; usually formed from cellulose acetate.


**FARADAY CAGE.** In the Everhart-Thornley detector, the wire mesh surrounding the scintillator, which is positively charged (40-200 V) to increase the collection of low-energy secondary electrons.

**FE.** See FIELD EMISSION SOURCE.

**FET.** See FIELD EFFECT TRANSISTOR.

**FIELD EFFECT TRANSISTOR (FET).** A component of the energy-dispersive spectrometer which is basically a preamplifier that integrates the total charge received from the detector and converts it into a proportional voltage signal that is carried into the linear amplifier.

**FIELD EMISSION SOURCE.** A very bright electron source capable of producing a fine beam; used in high-resolution SEMs.

**FILAMENT.** Cathodic electron source, synonymous with emitter; usually refers to tungsten hairpin filaments.

**FLUORESCENCE.** The excitation of lower-energy X-rays by higher-energy X-rays; synonymous with secondary emission.

**FLUORESCENT YIELD.** Interactions resulting in the emission of X-rays from a specimen; compare with ABSORPTION.

**FRAME C.** A quantitative EDS correction that performs background subtraction and resolves peak overlap.

**FULL WIDTH AT HALF MAXIMUM (FWHM).** A measure of spectrometer resolution expressed in terms of electron volts per channel; usually measured for the Mn K$\alpha$ peak from a radioactive Fe$^{55}$ source.

**FWHM.** See FULL WIDTH AT HALF MAXIMUM.


**GAMMA.** A signal-processing device that suppresses very dark or bright levels by intensifying the intermediate gray levels; synonymous with gamma modulation.

**GLOW DISCHARGE.** In sputtering, the luminescence of the plasma resulting from electrical discharge in the plasma.

**GRAY LEVELS.** Gradations of black to white in an image.

**GRID CAP.** See SHIELD.

**GUN CARTRIDGE.** Collectively, the filament and shield.


**HALATION, RESIDUAL.** Continued emission of light from a fluorescent CRT screen after the excitation has been removed.

**HIGH TENSION.** In the sputter coater, synonymous with ACCELERATING VOLTAGE.


**IMAGE INVERSION.** An SEM signal-processing device that reverses black and white.

**IMAGE PROCESSING.** Analytical image processing serves to extract specific data from an image (e.g., the number of inclusions per unit area); digital image processing enhances image clarity.

**INELASTIC COLLISIONS.** Electron-electron interactions that produce secondary electrons; cf. ELASTIC COLLISIONS. See also SECONDARY ELECTRONS.

**INFORMATION DEPTH.** See DEPTH OF PENETRATION; EXCITATION VOLUME.

**INTERFERENCE METHODS.** Double- or multiple-beam interference techniques used for very accurate measurement of thin-film thickness.

**ION BEAM SPUTTERING.** A method of producing high-resolution sputtered films using an ion beam focused on a metal target.

**k-FACTOR.** The ratio between the unknown and standard X-ray intensities used in quantitative analyses.

**KLM X-RAYS.** Collectively, X-rays resulting from electron transitions in the K, L, and M electron shells.

**$LaB_6$.** See LANTHANUM HEXABORIDE SOURCE.

**LANTHANUM HEXABORIDE ($LaB_6$) SOURCE.** An intense electron source used in high-resolution SEMs; cf. FIELD EMISSION SOURCE; TUNGSTEN FILAMENT.

**LATERAL-SHIFT METHOD.** A technique for recording low-magnification stereo pairs in which the specimen is moved sideways between photographs; cf. TILT METHOD.

**LIGHT PIPE.** A light-propagating medium between the scintillator and photomultiplier tube.

**LINE SCAN.** A method of X-ray spectrum display that reveals the relative distribution of an element by a vertical deflection of a line superimposed over an image. Synonymous with line profile.

**LITHIUM-DRIFTED SILICON DETECTOR [Si(Li) DETECTOR].** A solid-state (semiconductor) detector for X-rays composed of moderately pure silicon doped with lithium.

**MAGIC IV AND V.** Quantitative energy-dispersive spectroscopy programs.

**MAGNETRON SPUTTERING.** A coating method which eliminates thermal damage.

**MAGNIFICATION.** The specimen area irradiated by the electron beam. EMPTY MAGNIFICATION refers to increased magnification without improved resolution.

**MAP, X-RAY.** See X-RAY MAP.

**MASS ABSORPTION COEFFICIENT.** A function relating wavelength of the absorbed radiation and the atomic number of the absorbing element.

**MATRIX CORRECTIONS.** The quantitative conversion of peak intensity values into relative concentrations in energy-dispersive spectroscopy.

**MCA.** See MULTICHANNEL ANALYZER.

**MICROBALANCE, VACUUM.** A sensitive tool for indirectly measuring thin-film thickness as a function of weight during evaporation.

**MONOCHROMATOR.** A device used for the isolation of one wavelength of light in cathodoluminescence detectors.

**MONTE CARLO TECHNIQUES.** Calculation of the trajectory of incident electrons within a given matrix and the pathway of the X-rays generated during interaction.

**MOSELEY'S LAW.** A law stating that the square root of the frequency of the characteristic X-rays of the K, L, M, or N line series is directly proportional to atomic number; it predicts that sharp X-ray peaks will be observed and indicates that cathodoluminescence is a broad-band phenomenon.

**MULTICHANNEL ANALYZER (MCA).** An instrument that splits an input waveform into a number of channels with respect to a particular parameter of the input. In energy-dispersive spectroscopy, the MCA sorts pulses received from the amplifier, stores and retrieves spectra, and may also be capable of quantitative data reduction.

**NEGATIVE REPLICAS.** Single-stage impressions of a surface.

**NOISE.** Spurious background that interferes with data signals. See also SIGNAL-TO-NOISE RATIO.

**NONRADIATIVE TRANSITIONS.** The generation of heat phonons in a specimen during irradiation; cf. RADIATIVE TRANSITIONS.

**NUCLEUS.** In thin-film terminology, one of the initial sites that intercept and bind free metal atoms, gradually growing into islands that then coalesce and form a continuous thin film.

**NUMERICAL APERTURE.** An expression of the light-gathering power of a microscope objective lens.

**OBJECTIVE LENS.** The microscope lens controlling magnification and focus.

**OPTIMAL APERTURE.** In conventional photography, the lens opening giving the greatest depth of field.

**PARALLAX.** The horizontal displacement between the members of a stereo pair.

**PEAK OVERLAP.** Formation of a single peak when two closely spaced X-ray peaks cannot be resolved; the energy of the peak is the average of the characteristic energies of the original two peaks. See also FULL WIDTH AT HALF MAXIMUM.

**PELTIER-MODULE COLD STAGE.** A stage used in cool diode sputter coaters based upon the liberation of heat at one junction and the absorption of heat at the other junction when a current is passed around a circuit consisting of two different metals.

**PENNING SPUTTERING.** A form of ion beam sputtering conducted under high vacuum that produces extremely fine thin films.

**PHOTOMULTIPLIER TUBE (PMT).** A device that enhances very weak imaging signals by amplifying them through a cascade effect.

**PLASMA.** An ionized gas containing nearly equal numbers of electrons and positive ions; used in sputter coating.

**POLARIZED STEREO PROJECTION.** The simultaneous projection of a stereo pair through adjacent slide projectors onto a lenticular silver screen; cf. ANAGLYPHS.

**POSITIVE REPLICAS.** Two-stage replicas in which the first negative impression serves as a template for the second replica.

**PREAMPLIFIER.** See FIELD EFFECT TRANSISTOR.

**PRINCIPLE POINT.** The location of the stationary electron beam relative to the plane of the micrograph; used in stereo imaging.

**PULSE PILE-UP.** At excessively high count rates, the probability of two or more X-rays being simultaneously incident upon the detector; manifested as a distortion of peak symmetry or the formation of a SUM PEAK.

**QUALITATIVE ANALYSIS.** In energy-dispersive spectroscopy, identification of the elemental composition of a specimen.

**QUANTITATIVE ANALYSIS.** In energy-dispersive spectroscopy, determination of the relative concentrations of the detected elements.

**QUARTZ CRYSTAL OSCILLATOR.** An instrument used to measure the thickness of thin films.

**RADIATIVE TRANSITIONS.** The emission of electromagnetic radiation (electrons, photons, and X-rays) by an irradiated specimen; cf. NONRADIATIVE TRANSITIONS.

**RADIOFREQUENCY SPUTTERING.** Sputtering caused by the application of an alternating current to a target, inducing a radiofrequency voltage on the surface; the target is at the boundary of a plasma.

**REPLICAS.** Impressions of a surface. Negative replicas are single-stage impressions of a surface, and positive replicas are secondary impressions of a negative replica. See CELLULOSE ACETATE TAPE; DENTAL IMPRESSION MEDIA; EXTRACTION REPLICAS.

**RESISTANCE HEATING.** The induction of a temperature rise by passage of a current through a metal as in THERMAL EVAPORATION.

**RESOLUTION.** The ability to make the individual parts of an image distinguishable. Spatial (or point-to-point) resolution is the ability to separate two closely spaced dots in an image; in energy-dispersive spectroscopy, spatial resolution refers to the lateral dimensions of the X-ray volume. Spectral resolution is the ability to separate closely spaced peaks. See also FULL WIDTH AT HALF MAXIMUM.

**ROBINSON DETECTOR.** A backscattered electron detector originally designed for the observation of wet specimens.

**ROTARY EVAPORATION.** Thermal evaporation (from a point source) onto a specimen stage subjected to planetary motion (simultaneously rotating and tilting) in order to coat the specimen with a continuous thin film.

**ROTARY PUMP.** A mechanical pump that produces low vacuum ($\sim 10^{-2}$ torr) from atmospheric pressure by the physical displacement of air. Synonymous with mechanical, roughing, or low-vacuum pump.

**RUTHERFORD SCATTERING.** A general term for the process by which moving charged particles are scattered at various angles by interaction with the nuclei of atoms of a solid material; e.g., backscattered electrons undergo a Rutherford scattering.

**SATURATION.** The effective tungsten filament operating condition.

**SCAN GENERATOR.** A device used to simultaneously deflect the electron beam and the CRT display beam.

**SCAN RATE.** The speed at which the rastered beam passes over the specimen.

**SCAN ROTATION.** An electronic deflection of the electron beam that supplants specimen rotation.

**SCINTILLATOR.** A component of the electron detector which translates incident electron signals into a proportional number of photons that then propagate through the light pipe.

**SECONDARY ELECTRONS ($2°$ $e^-$).** The imaging signals arising from electron-electron (inelastic) collisions; cf. BACKSCATTERED ELECTRONS. See also EVERHART-THORNLEY DETECTOR.

**SECONDARY EMISSION.** See FLUORESCENCE.

**SELF-SHADOWING.** An artifact infrequently seen in thermal evaporation. Manifested as a prominent line running parallel to the shadowing direction, it is caused by shielding of the source by a prominent surface structure.

**SHADOWING.** The oblique deposition of an evaporant onto a stationary target used to enhance the fidelity of minute surface features. Synonymous with shadow casting.

**SHIELD.** A component of the electron gun surrounding the filament. Synonymous with Wehnelt cylinder and grid cap.

**SIGNAL PROCESSING.** Modulations of the imaging signal in the SEM. See also BLACK-LEVEL SUBTRACTION; GAMMA; IMAGE INVERSION; Y-MODULATION

**SIGNAL-TO-NOISE RATIO (SNR).** The relative value between the desired data signal and spurious fluctuations (noise) which reduce the quality of the data.

**Si(Li) DETECTOR.** See LITHIUM-DRIFTED SILICON DETECTOR.

**SINGLE-STAGE REPLICAS.** See NEGATIVE REPLICAS.

**SMOOTHING, SPECTRUM.** A method of X-ray spectrum manipulation that improves the signal-to-noise ratio by reducing the level of noise in an acquired spectrum.

**SNR.** See SIGNAL-TO-NOISE RATIO.

**SOLID-STATE DETECTOR.** Any semiconductor-based device, such as lithium-drifted silicon detectors and backscattered electron detectors.

**SPECTROSCOPY.** See ENERGY-DISPERSIVE SPECTROSCOPY; WAVELENGTH-DISPERSIVE SPECTROSCOPY.

**SPHERICAL ABERRATION.** An optical aberration that occurs when the electrons passing through the center of the magnetic field of a lens are focused in one plane and those at the perimeter of the beam, which travel at a different velocity, are focused in another plane; this problem is eliminated by the use of apertures.

**SPOT SIZE.** Diameter of the electron beam incident upon the specimen. Synonymous with BEAM DIAMETER.

**SPUTTER COATING.** A method of thin-film preparation based on the erosion of metal atoms from a target by an energetic gas plasma; compare with THERMAL EVAPORATION. See also COOL DIODE, DIODE, and TRIODE SPUTTER COATER; ION BEAM, MAGNETRON, PENNING, and RADIOFREQUENCY SPUTTERING.

**STEREO ANGLE.** In the tilt method of stereo recording, the angular difference between each member of the stereo pair.

**STEREO IMAGING.** The addition of a third dimension (depth) to two-dimensional images by recording two photographs of a given field of view at different orientations, and simultaneously viewing both photographs in a stereo viewer. See also ANAGLYPHS; LATERAL SHIFT METHOD; POLARIZED STEREO PROJECTION; TILT METHOD.

**STEREOPSIS.** Perception of a stereo image.

**STEREOSCOPY.** Quantitative spatial measurements of stereo pairs.

**STIGMATORS.** Weak lenses incorporated into the final lens of the SEM which exert a magnetic field having a magnitude equal to but opposite from that of asymmetric fields generated by the final lens. See also ASTIGMATISM.

**STRIPPING REPLICAS.** Single-stage cellulose acetate impressions of inert surfaces which remove and preserve adherent surface debris, oxides, or inclusions.

**STRIPPING, SPECTRUM.** A technique used when peak overlap obscures weak X-ray peaks. The presence of the lower-concentration element may be confirmed by stripping a spectrum of the more intense element.

**STROBOSCOPY.** Pulsation of the electron beam to reveal time-dependent responses, often used in cathodoluminescence studies.

**SUM PEAK.** An artifact encountered during pulse pile-up where two X-rays simultaneously entering the detector are counted as one X-ray, the energy of which is equal to the sum of both X-rays.

**TAKE-OFF ANGLE.** The angle between the sample surface and the line of sight to the center of the X-ray detector; a high take-off angle ($>30°$) diminishes absorption of X-rays by the specimen.

**THERMAL EVAPORATION.** A method of thin-film preparation conducted under high vacuum; when a current is passed through the target metal, it vaporizes, and the metal recondenses on all surfaces within a line of sight of the source metal. Synonymous with resistance heating; cf. SPUTTER COATING. See also ROTARY EVAPORATION; SHADOWING.

**THERMOCOUPLE.** An electrical circuit consisting of two different metals in which an electromotive force is produced when the two junctions are at different temperatures.

**THIN FILMS.** Metal films 50 to 400 Å thick prepared by sputter coating or thermal evaporation, their purpose being to increase the secondary electron yield of nonconductive specimens and thereby improve imaging; carbon thin films, which increase conductivity without influencing the secondary electron yield, are used for X-ray analysis or as preliminary coatings before metal evaporation.

**TILT.** The angle of the specimen relative to the axis of the electron beam; at $0°$ tilt the specimen is perpendicular to the beam axis.

**TILT METHOD.** A technique for recording stereo pairs wherein an angular shift is applied between the two recordings; cf. LATERAL SHIFT METHOD.

**TRANSITIONS, ELECTRON.** Transitions that occur when an excited atom ejects an inner-shell electron, creating a vacancy that is filled by an outer-shell electron. The outer-shell electron emits an X-ray of energy equal to the difference in energy between the two shells.

**TRIODE SPUTTER COATER.** A modified diode sputter coater which limits thermal damage by introducing a high-potential anode between the target (cathode) and specimen (anode at ground potential).

**TUNGSTEN FILAMENT.** A component of the electron gun consisting of a hairpin-shaped tungsten wire connected to the high-voltage supply by two electrodes; when heated to incandescence, the filament releases electrons which form the imaging beam.

**TURBOMOLECULAR PUMP.** A hydrocarbon-free, turbine-like pump that achieves high vacuum by the displacement of gas through a series of rotating blades and stationary slotted disks; cf. DIFFUSION PUMP.

**TWO-STAGE REPLICAS.** See POSITIVE REPLICAS.

**UNDERSATURATION.** Formation of a multiple electron source at very low filament currents.

**VACUUM.** Low vacuum refers to $\sim 10^{-2}$ torr, high vacuum to $10^{-2}$ to $10^{-7}$ torr, and ultrahigh vacuum to $10^{-8}$ to $10^{-11}$ torr.

**VAPORIZATION TEMPERATURE.** The characteristic temperature at which a material vaporizes. See also THERMAL EVAPORATION.

**VOLTAGE CONTRAST.** The imaging of opens and shorts by electrically biasing an electronic device in the SEM.

**WAVEFORM MONITOR.** A deflection of the line scan that reveals topographic relationships of the specimen surface; also used to evaluate filament saturation and electron gun alignment.

**WAVELENGTH (λ).** The distance, measured in the direction of the propagation of a wave, between two successive points that are characterized by the same phase of vibration; in SEM, wavelength decreases as accelerating voltage increases, which in turn enhances resolution.

**WAVELENGTH-DISPERSIVE SPECTROSCOPY.** A method of X-ray analysis that employs a crystal spectrometer to discriminate characteristic X-ray wavelengths; cf. ENERGY-DISPERSIVE SPECTROSCOPY.

**WEHNELT CYLINDER.** See SHIELD.

**WHITE RADIATION.** See CONTINUUM.

**WINDOWLESS DETECTOR.** A special type of detector used in energy-dispersive spectroscopy for the detection of elements having low atomic number; cf. BERYLLIUM WINDOW.

**WOBBLER.** A signal generator whose output voltage is varied through a set range; used to vary the current passing through the objective lens. Useful for lens alignment.

**WORKING DISTANCE.** In the SEM, the separation between the specimen and final lens as controlled by the Z-axis; in energy-dispersive spectroscopy, the distance between the specimen and X-ray detector, as controlled by both the lateral movement of the detector and the Z-axis of the microscope.

**X-RAY MAP.** A map made by feeding spectral information into the SEM that shows the origin of the X-ray signal as bright dots against a dark background; used to relate compositional to image data. Synonymous with DOT MAP.

**X-RAYS.** Electromagnetic radiation lying between ultraviolet radiation and gamma rays in the spectrum.

**Y-MODULATION.** A mode of signal processing that displays an image as a series of closely spaced lines that correspond to an intensity contour map of the surface.

**ZAF CORRECTIONS.** Quantitative X-ray programs that correct for atomic number $(Z)$, absorption $(A)$, and fluorescence $(F)$ effects in a matrix.

**Z-AXIS.** A translation of the specimen stage that controls the distance between the specimen and final lens. See also WORKING DISTANCE.

# Appendix B:
# Characteristic X-ray Energies and Absorption Edge Energies in keV

| Atomic number | K$\alpha$ | K$\beta$ | K$_{(ab)}$ | L$\alpha$ | L$\beta_1$ | L$\beta_2$ | L$\gamma_1$ | L$_{III(ab)}$ | L$_{II(ab)}$ | M$_{v(ab)}$ | M$\alpha$ |
|---|---|---|---|---|---|---|---|---|---|---|---|
| 1 H . . . . . | | | 0.014 | | | | | | | | |
| 2 He . . . . | | | 0.025 | | | | | | | | |
| 3 Li . . . . . | | | 0.055 | | | | | 0.003 | 0.003 | | |
| 4 Be. . . . . | | | 0.116 | | | | | 0.002 | 0.002 | | |
| 5 B. . . . . . | | | 0.192 | | | | | 0.003 | 0.003 | | |
| 6 C . . . . . | 0.277 | | 0.283 | | | | | 0.002 | 0.002 | | |
| 7 N . . . . . | 0.392 | | 0.400 | | | | | 0.008 | 0.008 | | |
| 8 O . . . . . | 0.523 | | 0.531 | | | | | 0.009 | 0.009 | | |
| 9 F. . . . . . | 0.677 | | 0.687 | | | | | | | | |
| 10 Ne . . . . | 0.848 | | 0.874 | | | | | 0.019 | 0.019 | | |
| 11 Na . . . . | 1.041 | 1.067 | 1.070 | | | | | 0.031 | 0.031 | | |
| 12 Mg. . . . | 1.253 | 1.295 | 1.303 | | | | | 0.050 | 0.050 | | |
| 13 Al. . . . . | 1.486 | 1.553 | 1.559 | | | | | 0.073 | 0.073 | | |
| 14 Si . . . . . | 1.739 | 1.829 | 1.842 | | | | | 0.102 | 0.103 | | |
| 15 P. . . . . . | 2.013 | 2.136 | 2.142 | | | | | 0.128 | 0.129 | | |
| 16 S. . . . . . | 2.307 | 2.464 | 2.470 | | | | | 0.163 | 0.164 | | |
| 17 Cl. . . . . | 2.621 | 2.815 | 2.820 | | | | | 0.197 | 0.199 | | |
| 18 Ar. . . . . | 2.957 | 3.190 | 3.200 | | | | | 0.224 | 0.246 | | |
| 19 K . . . . . | 3.312 | 3.589 | 3.609 | | | | | 0.295 | 0.298 | | |
| 20 Ca. . . . . | 3.690 | 4.012 | 4.038 | 0.341 | 0.345 | | 0.350 | 0.346 | 0.350 | | |
| 21 Sc. . . . . | 4.088 | 4.460 | 4.496 | 0.395 | 0.400 | | 0.407 | 0.403 | 0.411 | | |
| 22 Ti . . . . . | 4.508 | 4.931 | 4.964 | 0.452 | 0.458 | | 0.460 | 0.454 | 0.460 | | |
| 23 V. . . . . . | 4.949 | 5.426 | 5.464 | 0.511 | 0.519 | | 0.520 | 0.513 | 0.519 | | |
| 24 Cr. . . . . | 5.411 | 5.924 | 5.987 | 0.573 | 0.583 | | 0.583 | 0.574 | 0.581 | | |
| 25 Mn . . . . | 5.894 | 6.489 | 6.537 | 0.637 | 0.649 | | 0.652 | 0.641 | 0.650 | | |
| 26 Fe. . . . . | 6.398 | 7.057 | 7.111 | 0.705 | 0.718 | | 0.721 | 0.709 | 0.720 | | |
| 27 Co . . . . | 6.924 | 7.648 | 7.709 | 0.776 | 0.791 | | 0.794 | 0.779 | 0.794 | | |
| 28 Ni. . . . . | 7.471 | 8.263 | 8.331 | 0.851 | 0.869 | | | 0.854 | 0.871 | | |
| 29 Cu . . . . | 8.040 | 8.904 | 8.981 | 0.930 | 0.950 | | | 0.933 | 0.953 | | |
| 30 Zn. . . . . | 8.630 | 9.570 | 9.661 | 1.012 | 1.034 | | | 1.022 | 1.045 | | |
| 31 Ga . . . . | 9.241 | 10.262 | 10.395 | 1.098 | 1.125 | | | 1.144 | 1.171 | | |
| 32 Ge . . . . | 9.874 | 10.979 | 11.100 | 1.188 | 1.218 | | | 1.214 | 1.245 | | |
| 33 As. . . . | 10.530 | 11.722 | 11.867 | 1.282 | 1.317 | | | 1.324 | 1.359 | | |
| 34 Se. . . . | 11.207 | 12.492 | 12.656 | 1.379 | 1.419 | | | 1.434 | 1.475 | | |
| 35 Br. . . . | 11.907 | 13.287 | 13.470 | 1.480 | 1.526 | | | 1.552 | 1.599 | | |
| 36 Kr. . . . | 12.631 | 14.107 | 14.320 | 1.586 | 1.636 | | 1.703 | 1.674 | 1.729 | | |
| 37 Rb . . . | 13.373 | 14.956 | 15.201 | 1.694 | 1.752 | | 2.050 | 1.806 | 1.866 | | |
| 38 Sr. . . . | 14.140 | 15.829 | 16.106 | 1.806 | 1.871 | | 2.196 | 1.941 | 2.008 | | |
| 39 Y. . . . . | 14.931 | 16.731 | 17.037 | 1.922 | 1.995 | | 2.346 | 2.080 | 2.155 | | |
| 40 Zr. . . . | 15.744 | 17.660 | 17.995 | 2.042 | 2.124 | 2.219 | 2.302 | 2.221 | 2.305 | | |
| 41 Nb . . . | 16.581 | 18.614 | 18.989 | 2.166 | 2.257 | 2.367 | 2.461 | 2.374 | 2.467 | | |
| 42 Mo . . . | 17.441 | 19.600 | 20.002 | 2.293 | 2.394 | 2.518 | 2.623 | 2.524 | 2.629 | | |

| Atomic number | Kα | Kβ | K$_{(ab)}$ | Lα | Lβ$_1$ | Lβ$_2$ | Lγ$_1$ | L$_{III(ab)}$ | L$_{II(ab)}$ | M$_{v(ab)}$ | Mα |
|---|---|---|---|---|---|---|---|---|---|---|---|
| 43 Tc.... | 18.325 | 20.608 | 21.054 | 2.424 | 2.536 | 2.674 | 2.792 | 2.677 | 2.795 | | |
| 44 Ru ... | 19.233 | 21.646 | 22.117 | 2.558 | 2.683 | 2.835 | 2.964 | 2.838 | 2.966 | | |
| 45 Rh ... | 20.165 | 22.712 | 23.218 | 2.696 | 2.834 | 3.001 | 3.143 | 3.002 | 3.144 | | |
| 46 Pd.... | 21.121 | 23.806 | 24.349 | 2.838 | 2.990 | 3.171 | 3.328 | 3.172 | 3.329 | | |
| 47 Ag.... | 22.101 | 24.928 | 25.512 | 2.984 | 3.150 | 3.347 | 3.519 | 3.350 | 3.522 | | |
| 48 Cd ... | 23.106 | 26.081 | 26.711 | 3.133 | 3.316 | 3.528 | 3.716 | 3.538 | 3.727 | | |
| 49 In .... | 24.136 | 27.260 | 27.937 | 3.286 | 3.487 | 3.713 | 3.920 | 3.729 | 3.937 | | |
| 50 Sn.... | 25.191 | 28.467 | 29.200 | 3.443 | 3.662 | 3.904 | 4.130 | 3.928 | 4.156 | | |
| 51 Sb.... | 26.271 | 29.706 | 30.491 | 3.604 | 3.843 | 4.100 | 4.347 | 4.132 | 4.381 | | |
| 52 Te.... | 27.377 | 30.974 | 31.813 | 3.769 | 4.029 | 4.301 | 4.570 | 4.342 | 4.612 | | |
| 53 I ..... | 28.508 | 32.272 | 33.170 | 3.937 | 4.220 | 4.507 | 4.800 | 4.559 | 4.853 | | |
| 54 Xe.... | 29.666 | 33.599 | 34.551 | 4.109 | 4.422 | 4.720 | 5.036 | 4.783 | 5.103 | | |
| 55 Cs.... | 30.851 | 34.961 | 35.983 | 4.286 | 4.619 | 4.935 | 5.279 | 5.011 | 5.359 | | |
| 56 Ba.... | 32.062 | 36.354 | 37.443 | 4.465 | 4.827 | 5.156 | 5.530 | 5.247 | 5.624 | | |
| 57 La.... | 33.299 | 37.771 | 38.932 | 4.650 | 5.041 | 5.383 | 5.788 | 5.490 | 5.897 | | 0.833 |
| 58 Ce ... | 34.566 | 39.223 | 40.447 | 4.839 | 5.261 | 5.612 | 6.051 | 5.729 | 6.169 | | 0.883 |
| 59 Pr.... | 35.860 | 40.711 | 41.995 | 5.033 | 5.488 | 5.849 | 6.321 | 5.969 | 6.445 | | 0.929 |
| 60 Nd ... | 37.182 | 42.231 | 43.577 | 5.229 | 5.721 | 6.088 | 6.602 | 6.215 | 6.728 | | 0.978 |
| 61 Pm ... | 38.532 | 43.783 | 45.190 | 5.432 | 5.960 | 6.338 | 6.891 | 6.462 | 7.022 | | |
| 62 Sm ... | 39.911 | 45.366 | 46.835 | 5.635 | 6.204 | 6.586 | 7.177 | 6.720 | 7.340 | | 1.081 |
| 63 Eu ... | 41.320 | 46.987 | 48.498 | 5.845 | 6.455 | 6.842 | 7.479 | 6.981 | 7.633 | | 1.131 |
| 64 Gd ... | 42.757 | 48.642 | 50.228 | 6.056 | 6.712 | 7.102 | 7.784 | 7.243 | 7.930 | | 1.185 |
| 65 Tb.... | 44.226 | 50.325 | 51.983 | 6.272 | 6.977 | 7.365 | 8.100 | 7.515 | 8.252 | | 1.240 |
| 66 Dy ... | 45.724 | 52.058 | 53.840 | 6.494 | 7.246 | 7.634 | 8.417 | 7.796 | 8.588 | | 1.293 |
| 67 Ho ... | 47.253 | 53.813 | 55.600 | 6.719 | 7.524 | 7.910 | 8.746 | 8.067 | 8.912 | | 1.347 |
| 68 Er.... | 48.813 | 55.606 | 57.457 | 6.947 | 7.809 | 8.188 | 9.087 | 8.357 | 9.263 | | 1.405 |
| 69 Tm ... | 50.406 | 57.437 | 59.376 | 7.179 | 8.100 | 8.467 | 9.424 | 8.650 | 9.616 | | 1.462 |
| 70 Yb.... | 52.030 | 59.322 | 61.313 | 7.414 | 8.400 | 8.757 | 9.778 | 8.944 | 9.979 | | 1.521 |
| 71 Lu ... | 53.687 | 61.235 | 63.306 | 7.654 | 8.708 | 9.047 | 10.142 | 9.242 | 10.346 | | 1.581 |
| 72 Hf.... | 55.382 | 63.183 | 65.347 | 7.898 | 9.021 | 9.346 | 10.514 | 9.558 | 10.736 | | 1.644 |
| 73 Ta.... | 57.098 | 65.125 | 67.406 | 8.145 | 9.342 | 9.650 | 10.893 | 9.874 | 11.129 | | 1.709 |
| 74 W .... | 58.856 | 67.140 | 69.159 | 8.396 | 9.671 | 9.960 | 11.284 | 10.202 | 11.540 | 1.809 | 1.774 |
| 75 Re.... | 60.648 | 69.199 | 71.678 | 8.651 | 10.008 | 10.274 | 11.683 | 10.537 | 11.960 | | 1.842 |
| 76 Os.... | 62.477 | 71.298 | 73.866 | 8.910 | 10.354 | 10.597 | 12.093 | 10.866 | 12.380 | | 1.914 |
| 77 Ir.... | 64.339 | 73.438 | 76.108 | 9.174 | 10.706 | 10.919 | 12.510 | 11.210 | 12.819 | 2.041 | 1.978 |
| 78 Pt.... | 66.241 | 75.618 | 78.386 | 9.441 | 11.069 | 11.249 | 12.940 | 11.557 | 13.266 | 2.122 | 2.048 |
| 79 Au ... | 68.177 | 77.840 | 80.723 | 9.712 | 11.440 | 11.583 | 13.379 | 11.919 | 13.734 | 2.206 | 2.120 |
| 80 Hg ... | 70.154 | 80.103 | 83.113 | 9.987 | 11.821 | 11.922 | 13.828 | 12.286 | 14.213 | 2.295 | 2.195 |
| 81 Tl.... | 72.167 | 82.497 | 85.529 | 10.267 | 12.211 | 12.270 | 14.289 | 12.658 | 14.696 | 2.389 | 2.268 |
| 82 Pb.... | 74.221 | 84.859 | 88.014 | 10.550 | 12.612 | 12.621 | 14.762 | 13.039 | 15.205 | 2.484 | 2.342 |
| 83 Bi.... | 76.315 | 87.328 | 90.471 | 10.837 | 13.021 | 12.978 | 15.245 | 13.419 | 15.715 | 2.579 | 2.418 |
| 84 Po.... | 78.452 | 89.781 | 93.112 | 11.129 | 13.445 | 13.338 | 15.741 | 13.817 | 16.244 | | |
| 85 At.... | 80.624 | 92.287 | 95.740 | 11.425 | 13.874 | | 16.249 | 14.215 | 16.784 | | |
| 86 Rn ... | 82.843 | 94.850 | 98.418 | 11.725 | 14.313 | | 16.768 | 14.618 | 17.337 | | |
| 87 Fr.... | 85.110 | 97.460 | 101.147 | 12.029 | 14.768 | 14.448 | 17.300 | 15.028 | 17.904 | | |
| 88 Ra.... | 87.419 | 100.113 | 103.900 | 12.338 | 15.233 | 14.839 | 17.845 | 15.444 | 18.482 | | |
| 89 Ac.... | 89.773 | 102.829 | 106.759 | 12.650 | 15.710 | 15.227 | 18.405 | 15.865 | 19.078 | | |
| 90 Th.... | 92.174 | 105.591 | 109.630 | 12.967 | 16.199 | 15.621 | 18.979 | 16.299 | 19.696 | 3.332 | 2.991 |
| 91 Pa.... | 94.627 | 108.409 | 112.581 | 13.288 | 16.699 | 16.022 | 19.565 | 16.731 | 20.311 | | 3.077 |
| 92 U .... | 97.131 | 111.281 | 115.610 | 13.612 | 17.217 | 16.425 | 20.164 | 17.164 | 20.939 | 3.552 | 3.165 |
| 93 Np ... | 99.407 | 113.725 | 118.619 | 13.942 | 17.747 | 16.837 | 20.781 | 17.614 | 21.596 | | |
| 94 Pu... | 101.857 | 116.943 | 121.200 | 14.276 | 18.291 | 17.252 | 21.414 | 18.066 | 22.262 | | |
| 95 Am... | 104.431 | 120.350 | 124.876 | 14.615 | 18.849 | 17.673 | 22.061 | 18.525 | 22.944 | | |
| 96 Cm... | 107.139 | 122.733 | 128.088 | 14.953 | 19.399 | 18.096 | 22.703 | 18.990 | 23.640 | | |
| 97 Bk... | 109.991 | 126.490 | 131.357 | 15.304 | 19.961 | 18.529 | 23.389 | 19.461 | 24.352 | | |
| 98 Cf... | 112.999 | 127.794 | 134.683 | 15.652 | 20.557 | 18.983 | 24.070 | 19.938 | 25.080 | | |

# Appendix C:
# Manufacturers and Suppliers

## SEM SUPPLIES

BALZERS UNION
8 Sagamore Park Road
Hudson, NH 03051

EBTEC CORP.
120 Shoemaker Lane
Agawam, MA 01001

ELECTRON MICROSCOPY
SCIENCES
P.O. Box 251
Ft. Washington, PA 19034

ERNEST F. FULLAM, INC.
P.O. Box 444
Schenectady, NY 12301

LADD RESEARCH INDUSTRIES,
INC.
P.O. Box 1005
Burlington, VT 05402

TED PELLA, INC.
P.O. Box 510
Tustin, CA 92680

POLYSCIENCES
Paul Valley Industrial Park
Warrington, PA 18976

## SEM MANUFACTURERS

AMRAY INC.
160 Middlesex Turnpike
Bedford, MA 01730

BAUSCH & LOMB/ARL
9545 Wentworth Street
Sunland, CA 91040

CAMBRIDGE INSTRUMENTS,
INC.
40 Robert Pitt Drive
Monsey, NY 10952

ISI, INC.
3255 6C Scott Blvd.
Santa Clara, CA 95051

JEOL, INC.
11 Dearborn Road
Peabody, MA 01960

NSA HITACHI
460E Middlefield Road
Mt. View, CA 94043

PHILLIPS ELECTRONICS
INSTR., INC.
85 McKee Drive
Mahwah, NJ 07430

## EDS MANUFACTURERS

EDS INTERNATIONAL, INC.
103 Schelter Road
P.O. Box 135
Prairie View, IL 60069

EG & G ORTEC
100 Midland Road
Oak Ridge, TN 37830

KEVEX CORP.
1101 Chess Drive
Foster City, CA 94404

PGT, INC.
1200 State Street
Princeton, NJ 08540

TRACOR NORTHERN, INC.
2551 W. Beltline Hwy.
Middleton, WI 53562

# SPUTTER COATERS AND VACUUM BELL JARS

BALZERS UNION
8 Sagamore Park Road
Hudson, NH 03051

EDWARDS HIGH VACUUM INC.
133 Standard Street
El Segundo, CA 90245

ERNEST F. FULLAM, INC.
P.O. Box 444
Schenectady, NY 12301

LADD RESEARCH INDUSTRIES, INC.
P.O. Box 1005
Burlington, VT 05402

TED PELLA, INC.
P.O. Box 510
Tustin, CA 92680

POLARON INSTRUMENTS, INC.
2293 Amber Drive
Line Lexington Ind. Park
Hatfield, PA 19440

SEEVAC, INC.
683 Regency Drive
Pittsburgh, PA 15239

STRUCTURE PROBE, INC.
P.O. Box 342
West Chester, PA 19380

TOUSIMIS
P.O. Box 2189
Rockville, MD 20852

# METALLOGRAPHIC SUPPLIERS

BUEHLER LTD
2120 Greenwood
Evanston, Illinois 60204

LECO CORP.
3000 Lakeview Avenue
St. Joseph, MI 49085

MAGER SCIENTIFIC INC.
P.O. Box 160
Dexter, MI 48130

STRUERS INC.
20102 Progress Drive
Cleveland, OH 44136

# Index

NOTE: The symbol (F) or (T) indicates that information is presented in a figure or a table.